Complex Data Analytics with Formal Concept Analysis

Rokia Missaoui • Léonard Kwuida
Talel Abdessalem
Editors

Complex Data Analytics with Formal Concept Analysis

 Springer

Editors
Rokia Missaoui
University of Quebec in Outaouais
Gatineau, QC, Canada

Léonard Kwuida
Business School
Bern University of Applied Sciences
Bern, Switzerland

Talel Abdessalem
Place Marguerite Perey
Télécom-Paris, Institut Polytechnique
Palaiseau, France

ISBN 978-3-030-93280-0   ISBN 978-3-030-93278-7   (eBook)
https://doi.org/10.1007/978-3-030-93278-7

This Springer imprint is published by the registered company Springer Nature Switzerland AG
The registered company address is: Gewerbestrasse 11, 6330 Cham, Switzerland

In memory of Vincent Duquenne
27.06.1950–26.02.2020.

Foreword

It was in 1970s that Galois connections and respective lattices of closed sets (studied before in Mathematics by Garrett Birkhoff and Oystein Ore) were found useful for modeling information structures and processes [3,26]. After several decades of research in formal concept analysis (FCA) no one can say now that FCA proposes hardly scalable techniques for the analysis of binary data. Highly efficient FCA algorithms with various options for approximation strategies are now widely used for the analysis of complex voluminous heterogeneous data.

In spite of intrinsic complexity of computational problems related to unrestricted generation of both formal concepts and implication bases [15,16], several efficient FCA algorithms were found already around year 2000 [10,19] and new efficient implementations show excellent scalability [1]. Numerous approaches to partial generation of concepts and implications were proposed, based on interestingness constraints [17] and probabilistic considerations [2,5].

Models for treating complex data with FCA-based approaches are manifold. Several approaches were proposed and developed for relational data, first considered through the prism of conceptual scaling, which reduces complex data to binary (or unary, in terms of Rudolf Wille). Most popular approaches to treating complex data in FCA "directly", i.e., without binarizing (scaling, in FCA terms) them, are logical concept analysis [7], pattern structures [9], fuzzy concept analysis [4], relational concept analysis [11], triadic concept analysis [13,20,22,23], polyadic CA [25], and probabilistic FCA [14]. Recent interest in natural language processing, knowledge graphs, and social network analysis inspired development of new FCA-based approaches [6,8,12,21].

The recent wave of interest in deep neural networks is tempered by the problems of explainability and robustness of proposed solutions. For some applied domains, like medicine, law, and finance, these issues are crucial: experts would not accept efficient accurate solutions that do not provide acceptable explanations. FCA can propose a broad scope of tools for finding interpretable solutions, since explainability is in the core of FCA. Several attempts were made already to combine neural network efficiency with explainability provided by FCA-based approaches [18,24].

This volume presents an important step in all the above-mentioned directions: meeting the challenge of big and complex data, combining FCA-based approaches with methods based on neural networks to guarantee explainability of results.

Moscow, Russia   Sergei O. Kuznetsov
August 2020

References

1. Simon Andrews. A 'best-of-breed' approach for designing a fast algorithm for computing fixpoints of galois connections. *Information Sciences*, 295:633–649, 2015.
2. Albert Atserias, José L. Balcázar, and Marie Ely Piceno. Relative entailment among probabilistic implications. *Log. Methods Comput. Sci.*, 15(1), 2019.
3. Marc Barbut and Bernard Monjardet. *Ordre et classification: algèbre et combinatoire*. Hachette, Paris, 1970.
4. Radim Belohlávek. Lattices of fixed points of fuzzy galois connections. *Math. Log. Q.*, 47(1):111–116, 2001.
5. Daniel Borchmann, Tom Hanika, and Sergei Obiedkov. Probably approximately correct learning of horn envelopes from queries. *Discrete Applied Mathematics*, 273:30–42, 2020.
6. Sébastien Ferré and Peggy Cellier. Graph-fca: An extension of formal concept analysis to knowledge graphs. *Discret. Appl. Math.*, 273:81–102, 2020.
7. Sébastien Ferré and Olivier Ridoux. A logical generalization of formal concept analysis. In Bernhard Ganter and Guy W. Mineau, editors, *Conceptual Structures: Logical, Linguistic, and Computational Issues, 8th International Conference on Conceptual Structures, ICCS 2000, Darmstadt, Germany, August 14–18, 2000, Proceedings*, volume 1867 of *Lecture Notes in Computer Science*, pages 371–384. Springer, 2000.
8. Boris A. Galitsky, Dmitry I. Ilvovsky, and Sergey O. Kuznetsov. Detecting logical argumentation in text via communicative discourse tree. *J. Exp. Theor. Artif. Intell.*, 30(5):637–663, 2018.
9. Bernhard Ganter and Sergei O. Kuznetsov. Pattern structures and their projections. In Harry S. Delugach and Gerd Stumme, editors, *Conceptual Structures: Broadening the Base, 9th International Conference on Conceptual Structures, ICCS 2001, Stanford, CA, USA, July 30-August 3, 2001, Proceedings*, volume 2120 of *Lecture Notes in Computer Science*, pages 129–142. Springer, 2001.
10. Robert Godin, Rokia Missaoui, and Hassan Alaoui. Incremental concept formation algorithms based on galois (concept) lattices. *Computational Intelligence*, 11:246–267, 1995.
11. Mohamed Rouane Hacene, Marianne Huchard, Amedeo Napoli, and Petko Valtchev. Relational concept analysis: mining concept lattices from multi-relational data. *Ann. Math. Artif. Intell.*, 67(1):81–108, 2013.

12. Mohamed Hamza Ibrahim, Rokia Missaoui, and Abir Messaoudi. Detecting communities in social networks using concept interestingness. In Iosif-Viorel Onut, Andrew Jaramillo, Guy-Vincent Jourdan, Dorina C. Petriu, and Wang Chen, editors, *Proceedings of the 28th Annual International Conference on Computer Science and Software Engineering, CASCON 2018, Markham, Ontario, Canada, October 29–31, 2018*, pages 81–90. ACM, 2018.

13. Dmitry I. Ignatov, Dmitry V. Gnatyshak, Sergei O. Kuznetsov, and Boris G. Mirkin. Triadic formal concept analysis and triclustering: searching for optimal patterns. *Mach. Learn.*, 101(1–3):271–302, 2015.

14. Francesco Kriegel. Implications over probabilistic attributes. In Karell Bertet, Daniel Borchmann, Peggy Cellier, and Sébastien Ferré, editors, *Formal Concept Analysis - 14th International Conference, ICFCA 2017, Rennes, France, June 13–16, 2017, Proceedings*, volume 10308 of *Lecture Notes in Computer Science*, pages 168–183. Springer, 2017.

15. Sergei Kuznetsov. On the intractability of computing the duquenne-guigues base. *Journal of Universal Computer Science*, 10:927–933, 01 2004.

16. Sergei O. Kuznetsov. On computing the size of a lattice and related decision problems. *Order*, 18(4):313–321, 2001.

17. Sergei O. Kuznetsov and Tatiana P. Makhalova. On interestingness measures of formal concepts. *Information Sciences*, 442–443:202–219, 2018.

18. Sergei O. Kuznetsov, Nurtas Makhazhanov, and Maxim Ushakov. On neural network architecture based on concept lattices. In Marzena Kryszkiewicz, Annalisa Appice, Dominik Slezak, Henryk Rybinski, Andrzej Skowron, and Zbigniew W. Ras, editors, *Foundations of Intelligent Systems - 23rd International Symposium, ISMIS 2017, Warsaw, Poland, June 26–29, 2017, Proceedings*, volume 10352 of *Lecture Notes in Computer Science*, pages 653–663. Springer, 2017.

19. Sergei O. Kuznetsov and Sergei A. Obiedkov. Comparing performance of algorithms for generating concept lattices. *J. Exp. Theor. Artif. Intell.*, 14(2–3):189–216, 2002.

20. Fritz Lehmann and Rudolf Wille. A triadic approach to formal concept analysis. In *ICCS*, pages 32–43, 1995.

21. Rokia Missaoui, Sergei O. Kuznetsov, and Sergei A. Obiedkov, editors. *Formal Concept Analysis of Social Networks*. Lecture Notes in Social Networks. Springer, 2017.

22. Rokia Missaoui and Léonard Kwuida. Mining Triadic Association Rules from Ternary Relations. In *LNAI*, volume 6628, pages 204–218. 2011.

23. Rokia Missaoui, Pedro H. B. Ruas, Léonard Kwuida, and Mark A. J. Song. Pattern discovery in triadic contexts. In Mehwish Alam, Tanya Braun, and Bruno Yun, editors, *Ontologies and Concepts in Mind and Machine - 25th International Conference on Conceptual Structures, ICCS 2020, Bolzano, Italy, September 18–20, 2020, Proceedings*, volume 12277 of *Lecture Notes in Computer Science*, pages 117–131. Springer, 2020.

24. Sebastian Rudolph. Using FCA for encoding closure operators into neural networks. In Uta Priss, Simon Polovina, and Richard Hill, editors, *Conceptual Structures: Knowledge Architectures for Smart Applications, 15th International*

Conference on Conceptual Structures, ICCS 2007, Sheffield, UK, July 22–27, 2007, Proceedings, volume 4604 of *Lecture Notes in Computer Science*, pages 321–332. Springer, 2007.

25. George Voutsadakis. Polyadic concept analysis. *Order*, 19(3):295–304, 2002.
26. Rudolf Wille. Restructuring lattice theory: An approach based on hierarchies of concepts. In Ivan Rival, editor, *Ordered Sets*, volume 83 of *NATO Advanced Study Institutes Series*, pages 445–470. Springer Netherlands, 1982.

Preface

With the advent of complex and big data and the increasing number of studies towards their management and analysis, it becomes important to get a better insight into existing studies, trends, and challenges and rely on promising theories such as formal concept analysis (FCA) together with recently developed technologies to design new, accurate, and scalable solutions for big data analytics facilities.

FCA is an important formalism that is associated with a variety of research areas such as lattice theory, knowledge representation, data mining, machine learning, and semantic Web, to name a few. It is successfully exploited in an increasing number of application domains such as software engineering, information retrieval, social network analysis, and bioinformatics. The mathematical power of FCA comes from its concept lattice formalization in which each element captures a formal concept while the whole structure represents a hierarchy that offers conceptual clustering, browsing, and association rule mining.

Although there are significant theoretical and practical contributions within the FCA community, including the design and implementation of efficient algorithms and tools for concept lattice computation and exploitation, this book examines a set of important and relevant research directions in complex data management and updates the contribution of the FCA community for analyzing complex and large data. For example, formal concept analysis and some of its extensions are exploited, revisited, and coupled with recent processing paradigms to maximize the benefits in analyzing large data. This book is a follow-up project of the workshop BigFCA'2019—Formal Concept Analysis in the Big Data Era—which was jointly organized with the ICFCA'2019 Conference in Frankfurt (see https://icfca2019.frankfurt-university.de/bigfca.html).

This volume of eleven chapters is meant to cover the state of the art of the research on the intersection of FCA and complex data analysis in a more systematic and detailed manner than it was done in the workshop proceedings mentioned above.

Gatineau, QC, Canada
Bern, Switzerland
Palaiseau, France
September 2020

Rokia Missaoui
Léonard Kwuida
Talel Abdessalem

Acknowledgments

The production of this volume would not have been possible without the valuable involvement and efforts of the contributing authors and the following reviewers: Simon Andrews, Jaume Baixeries, Karell Bertet, Agnès Braud, Víctor Codocedo, Pablo Cordero, Miguel Couceiro, Christophe Demko, Manuel Enciso, Sébastien Ferré, Tom Hanika, Mohamed Ibrahim, Dmitry Ignatov, Mehdi Kaytoue, Léonard Kwuida, Florence Le Ber, Pierre Martin, Ángel Mora Bonilla, Emilio Muñoz-Velasco, Amedeo Napoli, Manuel Ojeda-Aciego, Domingo López-Rodríguez, and Uta Priss. We highly appreciate the efforts and commitment of all the authors and reviewers. Our warm thanks go to Sergei O. Kuznetsov who wrote the foreword. We also would like to express our gratitude to the Editorial Director Ronan Nugent and the Senior Editor Paul Drougas from Springer for their help in the preparation of this volume. Finally, the production of this volume was greatly facilitated thanks to two excellent and well-known platforms, namely *EasyChair* for organizing paper submission and review, and *Overleaf* as an efficient and user-friendly collaborative LaTeX editor.

Contents

8   Scalable Visual Analytics in FCA 167
Tim Pattison, Manuel Enciso, Ángel Mora, Pablo Cordero, Derek Weber,
and Michael Broughton

List of Contributors

Jaume Baixeries
Computer Science Department, Universitat Politècnica de Catalunya, Barcelona, Spain

Meryem Bendella
Aix-Marseille Université, Marseille, France

Karell Bertet
Laboratory L3i, La Rochelle University, La Rochelle, France

Salah Boukhetta
Laboratory L3i, La Rochelle University, La Rochelle, France

Agnès Braud
ICube UMR 7537, Universite de Strasbourg, CNRS, ENGEES, Strasbourg, France

Michael Broughton
Defence Science and Technology Group, Adelaide, SA, Australia

Víctor Codocedo
Departamento de Informática, Universidad Técnica Federico Santa María, Campus San Joaquin, Santiago de Chile, Chile

Pablo Cordero
Universidad de Málaga, Málaga, Spain

Miguel Couceiro
Université de Lorraine, CNRS, Inria, LORIA, Nancy, France

Christophe Demko
Laboratory L3i, La Rochelle University, La Rochelle, France

Xavier Dolques
ICube UMR 7537, Universite de Strasbourg, CNRS, ENGEES, Strasbourg, France

Dominik Dürrschnabel
Knowledge & Data Engineering Group, University of Kassel, Kassel, Germany
Interdisciplinary Research Center for Information System Design, University of
Kassel, Kassel, Germany

Dmitry Egurnov
National Research University Higher School of Economics, Moscow, Russia

Manuel Enciso
Universidad de Málaga, Málaga, Spain

Cyril Faucher
Laboratory L3i, La Rochelle University, La Rochelle, France

Sébastien Ferré
Univ Rennes, CNRS, IRISA, Rennes, France

Alain Gély
Université de Lorraine, CNRS, LORIA, Metz, France

Alain Gutierrez
LIRMM, Univ Montpellier, CNRS, Montpellier, France

Tom Hanika
Knowledge & Data Engineering Group, University of Kassel, Kassel, Germany
Interdisciplinary Research Center for Information System Design, University of
Kassel, Kassel, Germany

Marianne Huchard
LIRMM, Univ Montpellier, CNRS, Montpellier, France

Dmitry I. Ignatov
National Research University Higher School of Economics, Moscow, Russia

Mehdi Kaytoue
Infologic R&D, Bourg-Lès-Valence, France

Priscilla Keip
CIRAD, UPR AIDA, Montpellier, France
AIDA, Univ Montpellier, CIRAD, Montpellier, France

Léonard Kwuida
Business School, Bern University of Applied Sciences, Bern, Switzerland

Florence Le Ber
ICube UMR 7537, Universite de Strasbourg, CNRS, ENGEES, Strasbourg, France

Domingo López-Rodríguez
Universidad de Malaga, Departamento Matematica Aplicada, Malaga, Spain

Pierre Martin
CIRAD, UPR AIDA, Montpellier, France
AIDA, Univ Montpellier, CIRAD, Montpellier, France

Rokia Missaoui
University of Quebec in Outaouais, Gatineau, QC, Canada

Ángel Mora
Universidad de Málaga, Málaga, Spain

Emilio Muñoz-Velasco
Universidad de Malaga, Departamento Matematica Aplicada, Malaga, Spain

Amedeo Napoli
Université de Lorraine, CNRS, Inria, LORIA, Nancy, France

Cristina Nica
Nicolae Titulescu University of Bucharest, Bucharest, Romania

Manuel Ojeda-Aciego
Universidad de Malaga, Departamento Matematica Aplicada, Malaga, Spain

Tim Pattison
Defence Science and Technology Group, Adelaide, SA, Australia

Mohamed Quafafou
Aix-Marseille Université, Marseille, France

Jérémy Richard
Laboratory L3i, La Rochelle University, La Rochelle, France

Pierre Silvie
IRD, UMR IPME, Montpellier, France
CIRAD, UPR AIDA, Montpellier, France
AIDA, Univ Montpellier, CIRAD, Montpellier, France

Maximilian Stubbemann
Knowledge & Data Engineering Group, University of Kassel, Kassel, Germany
Interdisciplinary Research Center for Information System Design, University of
Kassel, Kassel, Germany
L3S Research Center, Leibniz University Hannover, Hannover, Germany

Dmitry Tochilkin
National Research University Higher School of Economics, Moscow, Russia

Derek Weber
Defence Science and Technology Group, Adelaide, SA, Australia

Acronyms

ACN Abstract Conceptual Navigation
FCA Formal Concept Analysis
FD Functional Dependency
IR Information Retrieval
KG Knowledge Graph
NLP Natural Language Processing
OA Object-Attribute
OAC Object-Attribute-Condition
PAC Probably Approximately Correct
PGP Projected Graph Pattern
RCA Relational Concept Analysis
TCA Triadic Concept Analysis

Chapter 1
Formal Concept Analysis and Extensions for Complex Data Analytics

Léonard Kwuida and Rokia Missaoui

1.1 Introduction

In this era of big data [15, 36, 43, 55], we are drowning not only in evolving data of various types but also in the generated patterns that are extracted from them. Although new software and hardware technologies have recently emerged to manage massive data, it is crucial to carefully design and implement new data mining and machine learning algorithms by either adapting existing solutions or developing new ones in order to discover relevant and unexpected patterns. The complexity of data in the form of images, graphs, texts as well as audio and video sequences generates many challenges for researchers and practitioners. In this volume, the contributing authors show how Formal Concept Analysis, which is a theoretically sound framework for knowledge representation and pattern extraction, can be exploited and/or expanded to analyze complex data and visualize the generated knowledge. As highlighted in [33], Formal Concept Analysis offers means to process dynamic, complex, and even uncertain, data when it is combined with other theories such as conceptual structures of Sowa, rough set theory, as well as classical, fuzzy or description logic. It has also been successfully used to get actionable patterns from data in many applications such as information retrieval, software reuse, Web and text mining, ontology engineering, and so on.

The main objective of the present volume is to get a better insight into existing studies, trends and challenges on complex data analytics, and identify promising theories such as Formal Concept Analysis (FCA) that have been recently exploited together with recent technologies (e.g., MapReduce and Spark) or theories (e.g., fuzzy logic) to design new, accurate and scalable solutions for complex data analysis.

L. Kwuida (✉)
Business School, Bern University of Applied Sciences, Bern, Switzerland
e-mail: leonard.kwuida@bfh.ch

R. Missaoui
University of Quebec in Outaouais, Gatineau, QC, Canada
e-mail: rokia.missaoui@uqo.ca

Mining complex data [54] refers to advanced methods and tools for mining and analyzing data with complex structures such as XML/Json data, text and image data, multidimensional data, graphs, sequences and streaming data.

This introductory chapter is organized as follows. Section 1.2 gives an overview about FCA as a conceptual clustering approach, and a theoretical basis for implication mining. In Sect. 1.3, extensions to this main theory are described while Sect. 1.4 gives a brief overview of complex data analysis in the literature. Section 1.5 announces the contributions inside this volume and closes this chapter.

1.2 Background

In this section, we recall key notions of FCA as a branch of applied mathematics, which is based on a formalization of concept and concept hierarchy [21, 22]. FCA allows also the computation of association rules and different types of implication bases through the notions of pseudo-intent [23] and minimal generators [49].

1.2.1 Formal Concepts and Line Diagrams

FCA is based on the notion of concept. A concept consists of two parts: an intent and an extent. The extent contains all objects belonging to the concept, and the intent all attributes common to all objects in the concept. To formalize this notion, the data are represented in a very basic type, called formal context. A formal context is a triple $\mathbb{K} := (G, M, I)$ where G is the set of objects, M the set of attributes, and I a binary relation stating that an object has an attribute. Table 1.1 shows a formal context with six transactions (100–600) and items contained in these transactions.

	Shirt	Jacket	Hiking Boots	Ski Pants	Shoes	Outerwear	Clothes	Footwear
100	×						×	
200		×	×			×	×	×
300			×	×		×	×	×
400					×			×
500					×			×
600		×				×	×	

Table 1.1: Transactions and items

For $A \subseteq G$ and $B \subseteq M$, the derivation operators are defined by:

$$A' := \{m \in M \mid aIm \text{ for all } a \in A\} \quad \text{and} \quad B' := \{g \in G \mid gIb \text{ for all } b \in B\}.$$

A formal concept is a pair (A,B) with $A \subseteq G$, $B \subseteq M$ such that $A' = B$ and $B' = A$. We call A the extent and B the intent of the formal concept (A,B). We denote by $\mathfrak{B}(\mathbb{K})$ (resp. $\mathrm{Ext}(\mathbb{K})$, $\mathrm{Int}(\mathbb{K})$) the set of formal concepts (resp. extents, intents) of $\mathbb{K}$.

In the context of Table 1.1, the pair $(\{200, 300, 600\}; \{Outerwear, Clothes\})$ is a concept, and displays a cluster of transactions containing items that are *Outerwear* and *Clothes*. Columns and rows can be rearranged so that the concept corresponds to a maximal rectangle full with crosses.

A subset X of objects (resp. attributes) is closed if $X'' = X$. Thus, intents and extents are closed subsets. Three main (equivalent) properties of the derivations are useful for understanding the structure of concept sets:

1. The pair $(',')$ forms a Galois connection between $\mathscr{P}(G)$ and $\mathscr{P}(M)$;
2. The mappings $c : X \mapsto X''$ (on $\mathscr{P}(G)$ and on $\mathscr{P}(M)$) are closure operators.
3. The set of closed subsets (of G or M) forms a complete lattice.

We recall that a Galois connection between two posets $(P, \leq)$ and $(Q, \leq)$ is a pair (φ, ψ) of antitone mappings $\varphi : P \to Q$ and $\psi : Q \to P$ such that $p \leq \psi\varphi p$ and $q \leq \varphi\psi q$ for all $p \in P$ and $q \in Q$. A closure operator on a poset $(P, \leq)$ is a map $c : P \to P$ such that $x \leq c(y) \iff c(x) \leq c(y)$, for all $x, y \in P$. Equivalently $x \leq c(x)$, $x \leq y \implies c(x) \leq c(y)$, and $cc(x) = c(x)$ for all $x, y \in P$. A lattice is a poset in which all finite subsets have a least upper bound and a greatest lower bound. A lattice is complete if every subset has a least upper bound and a greatest lower bound. For any subset X of L we denote by $\bigvee X$ its least upper bound and by $\bigwedge X$ its greatest lower bound, whenever they exist. For $X = \{x, y\}$ we write $x \vee y := \bigvee X$ and $x \wedge y := \bigwedge X$.

Thus, $(\mathfrak{B}(\mathbb{K}), \leq)$ is a complete lattice called a concept lattice of the context $\mathbb{K}$, when ordered by $(A_1, B_1) \leqslant (A_2, B_2) : \iff A_1 \subseteq A_2$ (or equivalently $B_1 \supseteq B_2$). This order formalizes the concept hierarchy, which states that a concept is more general if it contains more objects. Note that $(\mathfrak{B}(\mathbb{K}), \leq)$ is isomorphic to the complete lattices $(\mathrm{Ext}(\mathbb{K}), \subseteq)$ and $(\mathrm{Int}(\mathbb{K}), \supseteq)$. In fact, the basic theorem on concept lattices states that $\mathfrak{B}(G, M, I)$ is a complete lattice and that each complete lattice $(L, \leq)$ is isomorphic to $\mathfrak{B}(G, M, I)$ iff there are mappings $\gamma : G \to L$ and $\mu : M \to L$ such that γG (resp. μM) is $\bigvee$-dense (resp. $\bigwedge$-dense) in L, and $g I m \iff \gamma g \leq \mu m$. For $L = \mathfrak{B}(G, M, I)$ the mappings γ and μ are given by: $\gamma g = (g'', g')$ and $\mu m = (m', m'')$. The concepts γg are called object concepts and μm attributes concepts. They form the building blocks for reconstructing concepts. In fact, the equalities below hold for any concept (A, B):

$$\bigvee_{g \in A} \gamma g = (A, B) = \bigwedge_{m \in B} \mu m.$$

Finite concept lattices can be represented by labeled Hasse diagrams. Each node represents a concept. For a reduced labelling, we write g underneath of γg and m above μm. The extent of a concept, represented by a node a, is given by all labels in G from a downwards, and the intent by all labels in M from a upwards. Figure 1.1 shows the concept lattice of the context in Table 1.1. The node labelled by *Outerwear* represents the concept $(\{200, 300, 600\}; \{Outerwear, Clothes\})$.

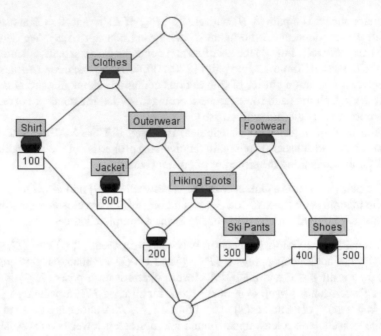

Fig. 1.1: A line diagram

Drawing a good diagram is a big challenge. Quite often, the size of the lattice is large and its structure is complex. Many techniques have been proposed: split the attributes set and get nested line diagrams, or decompose the lattice using congruences or tolerances to get factor or atlas representations. Other suggestions concentrate on the top of the lattice (iceberg lattices) or on zoom in/out using an is-a taxonomy on attributes (graded/generalized attributes). Visualization techniques are valuable tools for interpreting the extracted knowledge, and become more and more important, specially in the context of big and complex data.

1.2.2 Non Binary Data

Frequently, data are not directly encoded in a "binary" form. In this subsection we present two approaches used in FCA to handle non binary data.

1.2.2.1 Many-Valued Contexts

A many-valued context is a tuple (G, M, W, I) such that G is the set of objects, M the set of attribute names, W the set of attribute values, $I \subseteq G \times M \times W$ and every $m \in M$

is a partial map from G to W with $(g,m,w) \in I$ iff $m(g) = w$. Table 1.2 is an example of a many-valued context, with twelve objects and four attributes. Many-valued contexts

G / M	Gender	Age	Education	Salary	Target
1	Ma	Young	Higher	High	+
2	F	Middle	Special	High	+
3	F	Middle	Higher	Average	+
4	Ma	Old	Higher	High	+
5	Ma	Young	Higher	Low	−
6	F	Middle	Secondary	Average	−
7	F	Old	Special	Average	−
8	Ma	Old	Secondary	Low	−
9	F	Young	Special	High	τ
10	F	Old	Higher	Average	τ
11	Ma	Middle	Secondary	Low	τ
12	Ma	Old	Secondary	High	τ

Table 1.2: Many-valued classification context for credit scoring (borrowed from [29, 30])

can be transformed into binary contexts, via conceptual scaling. A conceptual scale for an attribute m of (G,M,W,I) is a binary context $\mathbb{S}_m := (G_m, M_m, I_m)$ such that $m(G) \subseteq G_m$. Intuitively, M_m discretizes or groups the attribute values into $m(G)$, and I_m describes how each attribute value $m(g)$ is related to the elements in M_m. For an attribute m of (G,M,W,I) and a conceptual scale $\mathbb{S}_m$ we derive a binary context $\mathbb{K}_m := (G, M_m, I^m)$ with $gI^m s_m : \iff m(g)I_m s_m$, where $s_m \in M_m$. This means that an object $g \in G$ is in relation with a scaled attribute s_m iff the value of m on g is in relation with s_m in $\mathbb{S}_m$. With a conceptual scale for each attribute we get the derived context $\mathbb{K}^S := (G, N, I^S)$ where $N := \bigcup\{M_m \mid m \in M\}$ and $gI^S s_m \iff m(g)I^m s_m$. In practice, the set of objects remains unchanged; each attribute name m is replaced with the scaled attributes $s_m \in M_m$. The choice of a suitable set of scales depends on the interpretation, and is usually done with the help of a domain expert. Nominal, ordinal, interordinal and dichotomic scales are examples of elementary scales [22]. A *Conceptual Information System* is a many-valued context together with a set of conceptual scales [45, 46]. Other scaling methods have also been proposed in [40–42]. The methods presented in [4, 34] to control the size of the attribute sets are actually a form of scaling.

An alternate way to handle such attributes without scaling is the use of pattern structures, described below.

1.2.2.2 Pattern Structures

Pattern structures consist of objects with descriptions that allow a semilattice operation on them, and arise naturally from ordered data [20]. The goal here is to process such data without prior scaling as described in Sect. 1.2.2.1. We consider a set of objects G and their descriptions (also called patterns) expressed by $(D, \sqcap)$. The operation $\sqcap$ is aimed at describing similarity between descriptions. A map $\delta : G \mapsto D$ assigns to each object its description. We assume that $(D, \sqcap)$ is a semilattice and that δG generates a complete subsemilattice of $(D, \sqcap)$. The two derivation operators $^{\square}$ are defined by:

$$A^{\square} := \sqcap\{\delta(g) \mid g \in A\} \text{ and } d^{\square} := \{g \mid d \sqsubseteq \delta(g)\} \quad \text{for } A \subseteq G \text{ and } d \in D,$$

where $\sqsubseteq$ is the subsemilattice order, i.e. $c \sqsubseteq d \iff c \sqcap d = c$. A pattern concept is a pair (A, d) with $A \subseteq G$, $d \in D$ such that $A^{\square} = d$ and $d^{\square} = A$.

The structure $(G, (D, \sqcap), \delta)$ is called a pattern context. The attributes set is defined by object descriptions $(D, \sqcap)$. A similar approach is pursued in logical concept analysis, where the set of attributes is replaced by logical formulae [17].

1.2.3 Implication Computation

We have seen that knowledge can be extracted from our basic information system (formal context) in terms of clusters, representing sets of objects described by the attributes. Another possibility is to extract rules between attributes. Let M be a set of properties or attributes. An implication between attributes in M is a pair (B_1, B_2) of subsets of M, and is usually denoted by $B_1 \to B_2$. It is a special case of association rules (also called approximate or partial implications) since the confidence is necessarily equal to 1. An implication $B_1 \to B_2$ holds in a context (G, M, I) if every object having all the attributes in B_1 also has all the attributes in B_2. This is equivalent to $B_2 \subseteq B_1''$. In the context of Table 1.1, the implication $\{Clothes, Footwear\} \to \{Hiking\ Boots, Outerwear\}$ holds.

Implications can be read from line diagrams, as follows:

$$B_1 \to B_2 \text{ holds in } (G, M, I) \iff \bigwedge\{\mu a \mid a \in B_1\} \le \mu m \text{ for all } m \in B_2.$$

A canonical direct basis associated with a context $\mathbb{K}$ is a concise set of implications of the form $A \to B \backslash A$ such that A is a minimal generator for $B \subseteq M$, i.e., a minimal subset such that its closure is equal to B. Other bases and rule sets have also been defined [21].

There is a natural correspondence between contexts, closure systems and implication systems. For each context, $\mathbb{K}$, we can find the set of implications, $\text{Imp}(\mathbb{K})$, that hold in $\mathbb{K}$. Conversely, given a set of implications, $\mathscr{L}$, we can construct a context $\mathbb{K}(\mathscr{L})$, such that $\text{Imp}(\mathbb{K}(\mathscr{L}))$ is the set of all implications that follow from $\mathscr{L}$. Therefore, extracting implications or concepts are both equivalent representations of the knowledge from a context.

1.3 Extensions to FCA

A few extensions have been proposed to the basic theory of FCA [33]. This includes Relational Concept Analysis [24], Logical Concept Analysis [17], Fuzzy FCA [2], rough FCA [39], graph-FCA [16], Triadic Concept Analysis [35], and more generally Polyadic Concept Analysis [52] and n-ary closed patterns [9].

1.3.1 Logical FCA

Logical Concept Analysis (LCA) [17, 18] is a generalization of FCA where attribute sets represent logical expressions. The power set of the attributes in the Galois connection is replaced with an arbitrary set of formulas as well as a deduction relation (i.e. subsumption), conjunctive and disjunctive operations.

In LCA, a context is a triple $(\mathcal{O}, \mathcal{L}, i)$ where $\mathcal{O}$ is a set of objects, $\langle \mathcal{L}; \models \rangle$ a lattice of formulas, and i a mapping from $\mathcal{O}$ to $\mathcal{L}$ which attaches a formula to an object. Given this context and the Galois connection (σ, τ) that holds between the two sets, a logical concept is a pair $c = (O, f)$ where $O \subseteq \mathcal{O}$ and f a formula in $\mathcal{L}$ such that $\sigma(O) \doteq f$ and $\tau(f) = O$.

1.3.2 Fuzzy FCA

Objects, attributes or incidences can be of fuzzy nature. Several sets of truth degrees have been considered, starting from the interval $[0, 1]$ [56] to lattices and residuated lattices. The main idea in is to uncover a Galois connection and deploy it for concept forming operators. A nice overview of fuzzy extensions of FCA is given in [3]. In [37] Ruiz-Calviño and Medina investigate the use of multilattices as underlying sets of truth-values. A multilattice is a poset $(P, \leq)$ in which for any finite subset X of P, each upper bound of X is above a minimal upper bound of X and each lower bound of X is below a maximal lower bound of X. The element x is below y or y is above x, whenever $x \leq y$. The existence of suprema and infima is relaxed to the existence of minimal upper bounds and maximal lower bounds.

1.3.3 Relational Concept Analysis

Relational concept analysis (RCA) was designed for mining formal concepts from multiple object sets described by both proper attributes and relations between objects [24, 27, 44].

Let $n, m \in \mathbb{N}$, then a relational context family (RCF) is a pair (K, R) such that

- $K := \{(G_i, M_i, I_i)\}_{1 \leq i \leq n}$ is a family of contexts with pairwise disjoint object sets,
- $R := \{r_k \subseteq G_i \times G_j\}_{1 \leq k \leq m}$ is a family of relations between the object sets of K.

Let G be the set of object sets of K. For any relation $r \subseteq G_i \times G_j$ in R, its domain and range are defined by $\mathrm{dom}(r) := G_i$ and $\mathrm{ran}(r) := G_j$. We will write $\mathbb{K}_i$ for (G_i, M_i, I_i) and $\mathbb{R}_k$ for (G_i, G_j, r_k). The family of relations starting from a context $\mathbb{K}_i$ is $\mathrm{rel}(\mathbb{K}_i) := \{r \in R \mid \mathrm{dom}(r) = G_i\}$. To process an RCF and take advantage from its relational structure, links between contexts/object sets will be turned into new attributes. This process is called relational scaling, and makes use of quantifiers such as universal, existential, at most, at least, and their combinations. Intuitively, the relational scaling of $\mathbb{K}_i$ along $r \in \mathrm{rel}(\mathbb{K}_i)$ extends M_i with new attributes and completes I_i accordingly. A scale for $\mathbb{K}_i$ with a quantifier Q along $r \subseteq G_i \times G_j$ with respect to $\mathfrak{B}(\mathbb{K}_j)$ is a context $\mathbb{K}_i^+ := (G_i, M_i^+, I_i^+)$, with $M_i^+ := \{Qrc \mid c \in \mathfrak{B}(\mathbb{K}_j)\}$ where Qrc is the name of the new (relational) attribute attached to $c \in \mathfrak{B}(\mathbb{K}_j)$.

If $Q = \exists$, then an object $g \in G_i$ will get the attribute $\exists rc$ in $\mathbb{K}_i^+$ if $r(g) \cap \mathrm{ext}(c) \neq \emptyset$ with $r(g) := \{h \in G_j \mid (g, h) \in r\}$. This is called existential scaling. The universal scaling is obtained with $Q = \forall$, and assigns the attribute $\forall rc$ to an object $g \in G_i$ in $\mathbb{K}_i^+$ if $r(g) \subseteq \mathrm{ext}(c)$. Usually $r(g)$ is required to be non empty. Each context $\mathbb{K}_i$ is then scaled over all relations r with $\mathrm{dom}(r) = G_i$ and gives its complete relational extension, via apposition. It contains $\mathfrak{B}(\mathbb{K}_i)$ as a subposet. The complete relational extension of an RCF, (K, R), is the family of all complete relational extensions of the contexts $\mathbb{K}_i \in K$. Relational scaling can help gain additional conceptual knowledge from sets of contexts [24].

1.3.4 Triadic Concept Analysis

Multidimensional data are ubiquitous in many real-life applications and Web resources. This is the case of multidimensional social networks, social resource sharing systems, and security policies. For instance, in the latter application, a user is authorized to use resources with given privileges under constrained conditions.

Triadic Concept Analysis (TCA) was originally introduced by Lehmann and Wille [35, 53] as an extension to (dyadic) FCA, to analyze data described by three sets K_1 (objects), K_2 (attributes) and K_3 (conditions) together with a ternary relation $Y \subseteq K_1 \times K_2 \times K_3$. Then $\mathbb{K} := (K_1, K_2, K_3, Y)$ is called a triadic context. With $(a_1, a_2, a_3) \in Y$, we mean that the object a_1 has the attribute a_2 under the condition a_3. Note that for a dyadic context (G, M, I) a concept is a pair (A, B) with $A \subseteq G$, $B \subseteq M$ and $A \times B \subseteq I$ and is maximal w.r.t inclusion in I. Similarly, a triadic concept of a triadic context $\mathbb{K}$ is a triple (A_1, A_2, A_3) (also denoted by $A_1 \times A_2 \times A_3$) with $A_1 \subseteq K_1$, $A_2 \subseteq K_2$, $A_3 \subseteq K_3$ and $A_1 \times A_2 \times A_3 \subseteq Y$ is maximal with respect to inclusion in Y. We then call A_1 the extent, A_2 the intent, and A_3 the modus of the triadic concept (A_1, A_2, A_3).

The derivation operators are introduced in a similar way. Let $\{i, j, k\} = \{1, 2, 3\}$ with $j < k$, $X_i \subseteq K_i$ and $(X_j, X_k) \subseteq K_j \times K_k$.[1] The $^{(i)}$-derivation [35] is defined by:

$$X_i^{(i)} := \{(a_j, a_k) \in K_j \times K_k \mid (a_i, a_j, a_k) \in Y \; \forall a_i \in X_i\} \tag{1.1}$$

$$(X_j, X_k)^{(i)} := \{a_i \in K_i \mid (a_i, a_j, a_k) \in Y \text{ for all } (a_j, a_k) \in X_j \times X_k\}. \tag{1.2}$$

The triadic concepts can be ordered, and form a complete trilattice [7, 35].

Triadic Concept Analysis was generalized by Voutsadakis [52] to Polyadic Concept Analysis to compute n-adic concepts and complete n-lattices from an n-adic context of size $n \geq 2$.

1.3.5 Approximation

The number of concepts and or rules can be extremely large. To handle the size of the set of extracted patterns many extensions have been proposed: fault-tolerant patterns [6], α-Galois lattices [51], graded attributes [4], generalized attributes [34], iceberg lattices [47], frequent itemsets and associations rules [57] with predefined thresholds for quality measures such as the support and the confidence.

Another way to cope with the large number of generated concepts from large datasets is to select only relevant ones from the concept set or the concept lattice [32]. It can be categorized into three groups: (i) formal context reduction using methods such as matrix factorization [11, 26], (ii) background knowledge (e.g., taxonomy or weight on attributes) or constraint consideration, and (iii) concept selection. The idea of the selection technique is to pick out a subset of concepts, objects or attributes based on relevancy measures. Several relevancy measures have been presented in the FCA framework such as concept stability, separation, probability, robustness, predictability Belohlávek and Trnecka [5], and Monocle [50]. Each selection measure aims at exploiting important features inside a formal context to capture concept relevancy. It includes the Galois connection between the sets of objects and attributes, the closure of an object/attribute subset, and the derivation operations. As such, the stability and separation [31], which depend primarily on these properties, have recently been introduced as the most prominent indices for assessing the concept quality [32]. The stability of a concept whose approximation was proposed in [1, 28] quantifies how its intent depends on the set of objects in its extent by validating either the derivation of the extent or the closure of the intent. The separation of a concept c represents the proportion of common cells in c between the area covered by the attributes of each object in its extent and the objects of every attribute in its intent.

OAC-triclusters are another kind of approximations of triadic concepts in a similar way as OA-biclusters are approximations of formal concepts. Theorems proving these approximation properties can be found in [12–14].

---

[1] We write $(X_j, X_k) \subseteq K_j \times K_k$ to mean that $X_j \subseteq K_j$ and $X_k \subseteq K_k$.

Yet another form of approximation in FCA is the one called rough FCA, which exploits rough set theory [39] that deals with uncertain data and imperfect knowledge. It helps find the interval in which a "vague concept" (i.e., not necessarily a concept) lies between two actual formal concepts named lower and upper approximations. The extent of the lower approximation contains all the objects of the vague concept while the upper approximation includes all objects which possibly belong to the concept.

1.4 Complex Data Analytics

In many application domains like engineering, biology, medicine, social media and economy, data have a rich but complex structure such as semi-structured data expressed in XML or Json format, texts, images, graphs, sequences, trees, multidimensional and streaming data. Complex data analysis aims at uncovering actionable patterns and learning models from such data in order to extract valuable insights from it. For example, graph mining seeks frequent common sub-graphs in graphs while sequence mining looks for frequent patterns in sequences with maximal common or prefix sub-sequences.

Since the last decade, researchers in data mining area surveyed existing work on complex data mining and analysis and pointed out the challenges that can be faced when analyzing them [10, 25].

In [36], the authors highlight the challenges of analyzing big data using machine learning techniques by focusing on each one of the main V's such as volume, velocity, veracity and variety. For the latter property mainly with respect to data type diversity, new learning algorithms such as deep, transfer and lifelong learning are considered as good ways to get insights, decisions and predictions from data.

In the very popular textbook on data mining techniques [25], Han et al. enumerate a set of complex data and their mining. For instance, multimedia data mining is an interdisciplinary field defined as the discovery of interesting patterns from image, video, audio, sequence or hypertext data. As an illustration, image mining relies on image processing, pattern recognition, machine learning, and data mining techniques to find image clusters or assign an image to a predefined class using raw and descriptive data.

In [48], the authors study important data mining topics in complex networks. To that end, they defined the notion of *Heterogeneous Information Networks* that can be perceived as a general framework to express complex networks with multiple types of nodes and links between them. The research topics covered are rank-based clustering and classification, meta-path-based similarity search/mining, and relationship prediction.

With respect to graph mining, a scalable and versatile graph mining system named *PegasusN* is presented in [38]. Such a system runs in a distributed environment on Hadoop and Spark, and is able to discover patterns and anomalies in massive graphs thanks to the efficient algorithms it integrates for different graph mining operations like graph structure analysis, subgraph identification as well as graph computation and visualization.

Data stream mining [19] is the process of discovering patterns from a stream which is an ordered sequence of continuous rapid data instances (e.g., credit card transactions, clickstream records, stock market data). This process has three features: (i) an infinite flow of large amount of data without any knowledge about the entire dataset, (ii) data evolves over time, and (iii) once the data is analyzed, it is either summarized or discarded but not stored. The book [8] by Bifet et al. contains a large set of machine learning algorithms and techniques that are presently used for mining data streams. The main algorithms are based either on classification, regression, clustering, or frequent pattern mining. Practical examples are also given using the open-source project called MOA.

1.5 Contributions

In addition to this introductory paper, the present volume contains ten contributions.

The contribution "Conceptual Navigation in Large Knowledge Graphs" of Sébastien Ferré proposes an exploration and querying paradigm named Abstract Conceptual Navigation (ACN) which merges querying and navigation to reconcile expressivity, usability, and scalability. The navigation space of ACN is a concept lattice. The ACN approach is applied to knowledge graphs (Graph-ACN) by relying on Graph-FCA, an extension of FCA to knowledge graphs. Graph-ACN can be efficiently implemented on top of SPARQL endpoints, and its expressivity can be increased in a modular way. Sparklis, an implementation available online, and a few application cases on large knowledge graphs, are presented.

In the article titled "FCA2VEC: Embedding Techniques for Formal Concept Analysis", Dominik Dürrschnabel, Tom Hanika, and Maximilian Stubbemann study a possible combination of modern embedding techniques with FCA. The authors enable the application of FCA notions to large data sets. It is shown how the cover relation of a concept lattice can be retrieved from a computationally feasible embedding. An improvement of the classical *node2vec* approach in low dimension is shown. For both directions, the overall requirement of FCA for explainable results is retained. The proposed techniques are evaluated on several datasets.

Under the title "Complex and heterogeneous data analysis using FCA and First order predicates", Karell Bertet, Christophe Demko, Salah Boukhetta, Jérémy Richard, and Cyril Faucher show how their algorithm called *NextPriorityConcept* can be used for the analysis of Boolean, categorized, numerical, character string and sequential data. Attributes are monadic predicates for each kind of data. A generic implementation of this algorithm in a development platform called GALACTIC allows the integration of new plugins corresponding to new data types.

In the paper titled "Computing Dependencies using FCA" by Jaume Baixeries, Victor Codocedo, Mehdi Kaytoue, and Amedeo Napoli, the authors use FCA and pattern structures to characterize and compute various kinds of constraints. The proposed unified framework can embrace a big diversity of constraints that define respective dependencies in data. The authors make a review of previously obtained results, discuss limitations and drawbacks of their approach and suggest further possible research directions.

In the paper "Patterns and Word Embeddings Based Query Expansion for Enhanced Microblogs Retrieval" by Meryem Bendella and Mohamed Quafafou, the defined work aims at improving tweets retrieval quality by proposing an FCA-based method of query expansion. The method uses word embeddings to enrich the patterns by adding similar words. The final query is given by merging the initial query with the extended one. The proposed method was evaluated on the TREC 2011 dataset.

In the paper "Dealing with Large Volumes of Complex Relational Data using RCA" by Agnès Braud, Xavier Dolques, Alain Gutierrez, Marianne Huchard, Priscilla Keip, Florence Le Ber, Pierre Martin, Cristina Nica and Pierre Silvie, the authors study how the issues raised by complex data can be managed through RCA. Using two datasets, one about the quality monitoring of waterbodies in France and the other about the use of pesticidal and antimicrobial plants in Africa, they study the limitations of different FCA algorithms, and their current implementations to explore these datasets with RCA. It is stated that RCA results can be summarized into a hierarchy of graph-patterns for helping the analysis by the domain expert. An example is given on temporal data.

In the paper "Scalable Visual Analytics in FCA" by Tim Pattison, Manuel Enciso, Angel Mora, Pablo Cordero, Derek Weber, and Michael Broughton, a visual analytic approach to FCA is developed by combining computational analysis with interactive visualization. The authors consider potential solutions to scalability challenges of FCA posed by big data: the time required to enumerate the vertices, edges and labels of the lattice digraph; the difficulty of responsive presentation of, and meaningful user interaction with a large digraph; the time required to enumerate (a basis for) all valid implications; and the discovery of insightful implications.

In their contribution titled "Formal methods in FCA and Big Data", Domingo Lopez-Rodrıguez, Emilio Munoz-Velasco, and Manuel Ojeda-Aciego survey the theoretical and technical foundations of some trends in FCA and present a summary of promising theoretical and practical applications of FCA that could be used to solve the problem of dealing with big data such as computing and using implicational systems.

In the paper "Towards Distributivity in FCA for Phylogenetic Data" by Alain Gély, Miguel Couceiro and Amedeo Napoli, the authors investigate the following problem: given a semilattice L obtained from a lattice by deleting the bottom element, is there a minimum distributive $\vee$-semilattice Ld such that L can be order embedded into Ld?

In the paper "Triclustering in Big Data Setting" by Dmitry Egurnov, Dmitry I. Ignatov, and Dmitry Tochilkin, the authors describe triclustering algorithms adapted for efficient calculations in distributed environments with the MapReduce model or parallelization mechanism provided by modern programming languages. OAC-triclustering provides a variant of approximation for formal triconcepts and thus continues development of OA-biclustering as another fault-tolerant approximation of formal concepts in dyadic case.[2] It is also demonstrated that OAC-family of algorithms can be efficiently parallelized by the independent processing of triples of the triadic formal context.

Acknowledgements The second author acknowledges the financial support of the Natural Sciences and Engineering Research Council of Canada (NSERC).

References

1. Babin, M.A., Kuznetsov, S.O.: Approximating concept stability. In: Formal Concept Analysis - 10th International Conference, ICFCA 2012, Leuven, Belgium, May 7–10, 2012. Proceedings. pp. 7–15 (2012)
2. Belohlávek, R.: Concept lattices and order in fuzzy logic. Ann. Pure Appl. Log. **128**(1–3), 277–298 (2004)
3. Belohlávek, R.: What is a fuzzy concept lattice? II. In: Kuznetsov, S.O., Slezak, D., Hepting, D.H., Mirkin, B.G. (eds.) Rough Sets, Fuzzy Sets, Data Mining and Granular Computing - 13th International Conference, RSFDGrC 2011, Moscow, Russia, June 25–27, 2011. Proceedings. Lecture Notes in Computer Science, vol. 6743, pp. 19–26. Springer (2011). https://doi.org/10.1007/978-3-642-21881-1_4
4. Belohlavek, R., De Baets, B., Konecny, J.: Granularity of attributes in formal concept analysis. Information Sciences **260**, 149–170 (2014)
5. Belohlávek, R., Trnecka, M.: Basic level in formal concept analysis: Interesting concepts and psychological ramifications. In: IJCAI 2013, Proceedings of the 23rd International Joint Conference on Artificial Intelligence, Beijing, China, August 3–9, 2013. pp. 1233–1239 (2013)
6. Besson, J., Pensa, R.G., Robardet, C., Boulicaut, J.F.: Constraint-based mining of fault-tolerant patterns from boolean data. In: International Workshop on Knowledge Discovery in Inductive Databases. pp. 55–71. Springer (2005)
7. Biedermann, K.: How triadic diagrams represent conceptual structures. In: ICCS. pp. 304–317 (1997)
8. Bifet, A., Gavaldà, R., Holmes, G., Pfahringer, B.: Machine Learning for Data Streams with Practical Examples in MOA. MIT Press (2018)
9. Cerf, L., Besson, J., Robardet, C., Boulicaut, J.F.: Closed patterns meet n-ary relations. TKDD **3**(1) (2009)
10. Cordeiro, R.L.F., Faloutsos, C., Jr., C.T.: Data Mining in Large Sets of Complex Data. Springer Briefs in Computer Science, Springer (2013)
11. Dias, S.M., Vieira, N.J.: A methodology for analysis of concept lattice reduction. Inf. Sci. **396**, 202–217 (2017)
12. Dmitry I. Ignatov, Dmitry V. Gnatyshak, Sergei O. Kuznetsov, and Boris Mirkin. Triadic formal concept analysis and triclustering: searching for optimal patterns. *Machine Learning*, pages 1–32, 2015.
13. Dmitry I. Ignatov, Sergei O. Kuznetsov, Jonas Poelmans, and Leonid E. Zhukov. Can triconcepts become triclusters? *International Journal of General Systems*, 42(6):572–593, 2013.
14. Dmitry I. Ignatov, Sergei O. Kuznetsov, and Jonas Poelmans. Concept-based biclustering for internet advertisement. In *ICDM Workshops*, pages 123–130. IEEE Computer Society, 2012.
15. Fan, W., Bifet, A.: Mining big data: current status, and forecast to the future. SIGKDD Explorations **14**(2), 1–5 (2012)
16. Ferré, S., Cellier, P.: Graph-fca: An extension of formal concept analysis to knowledge graphs. Discret. Appl. Math. **273**, 81–102 (2020)

---

[2] OAC stands for Object-Attribute-Condition, while OA is a shorthand for Object-Attribute.

17. Ferré, S., Ridoux, O.: A logical generalization of formal concept analysis. In: International Conference on Conceptual Structures: Logical, Linguistic, and Computational Issues. pp. 371–384. Springer (2000)
18. Ferré, S., Ridoux, O.: Introduction to logical information systems. Inf. Process. Manag. **40**(3), 383–419 (2004)
19. Gaber, M.M.: Advances in data stream mining. Wiley Interdiscip. Rev. Data Min. Knowl. Discov. **2**(1), 79–85 (2012)
20. Ganter, B., Kuznetsov, S.O.: Pattern structures and their projections. In: International conference on conceptual structures. pp. 129–142. Springer (2001)
21. Ganter, B., Obiedkov, S.A.: Conceptual Exploration. Springer (2016)
22. Ganter, B., Wille, R.: Formal Concept Analysis: Mathematical Foundations. Springer-Verlag New York, Inc. (1999), translator-C. Franzke
23. Guigues, J.L., Duquenne, V.: Familles minimales d'implications informatives résultant d'un tableau de données binaires. Mathématiques et Sciences humaines **95**, 5–18 (1986)
24. Hacene, M.R., Huchard, M., Napoli, A., Valtchev, P.: Relational concept analysis: mining concept lattices from multi-relational data. Ann. Math. Artif. Intell. **67**(1), 81–108 (2013)
25. Han, J., Kamber, M., Pei, J.: Data Mining: Concepts and Techniques, 3rd edition. Morgan Kaufmann (2011)
26. Horak, Z., Kudelka, M., Snásel, V.: Properties of concept lattice reduction based on matrix factorization. In: IEEE International Conference on Systems, Man, and Cybernetics, Manchester, SMC 2013, United Kingdom, October 13–16, 2013. pp. 333–338. IEEE (2013)
27. Huchard, M., Hacene, M.R., Roume, C., Valtchev, P.: Relational concept discovery in structured datasets. Annals of Mathematics and Artificial Intelligence **49**(1–4), 39–76 (2007)
28. Ibrahim, M.H., Missaoui, R.: Approximating concept stability using variance reduction techniques. Discret. Appl. Math. **273**, 117–135 (2020)
29. Ignatov, D.I.: Introduction to formal concept analysis and its applications in information retrieval and related fields. In: Braslavski, P., Karpov, N., Worring, M., Volkovich, Y., Ignatov, D.I. (eds.) Information Retrieval - 8th Russian Summer School, RuSSIR 2014, Nizhniy, Novgorod, Russia, August 18–22, 2014, Revised Selected Papers. Communications in Computer and Information Science, vol. 505, pp. 42–141. Springer (2014). https://doi.org/10.1007/978-3-319-25485-2_3
30. Ignatov, D.I., Kwuida, L.: Interpretable concept-based classification with shapley values. In: Alam, M., Braun, T., Yun, B. (eds.) Ontologies and Concepts in Mind and Machine - 25th International Conference on Conceptual Structures, ICCS 2020, Bolzano, Italy, September 18–20, 2020, Proceedings. Lecture Notes in Computer Science, vol. 12277, pp. 90–102. Springer (2020). https://doi.org/10.1007/978-3-030-57855-8_7
31. Klimushkin, M., Obiedkov, S.A., Roth, C.: Approaches to the selection of relevant concepts in the case of noisy data. In: ICFCA. vol. 20, pp. 255–266. Springer (2010)
32. Kuznetsov, S.O., Makhalova, T.P.: On interestingness measures of formal concepts. Inf. Sci. **442–443**, 202–219 (2018)
33. Kuznetsov, S.O., Poelmans, J.: Knowledge representation and processing with formal concept analysis. Wiley Interdiscip. Rev. Data Min. Knowl. Discov. **3**(3), 200–215 (2013)
34. Kwuida, L., Missaoui, R., Balamane, A., Vaillancourt, J.: Generalized pattern extraction from concept lattices. Annals of Mathematics and Artificial Intelligence **72**(1–2), 151–168 (2014)
35. Lehmann, F., Wille, R.: A Triadic Approach to Formal Concept Analysis. In: Proceedings of the Third International Conference on Conceptual Structures: Applications, Implementation and Theory. pp. 32–43 (1995)
36. L'Heureux, A., Grolinger, K., ElYamany, H.F., Capretz, M.A.M.: Machine learning with big data: Challenges and approaches. IEEE Access **5**, 7776–7797 (2017)
37. Medina, J., Ruiz-Calviño, J.: Fuzzy formal concept analysis via multilattices: First prospects and results. In: Szathmary, L., Priss, U. (eds.) Proceedings of The Ninth International Conference on Concept Lattices and Their Applications, Fuengirola (Málaga), Spain, October 11–14, 2012. CEUR Workshop Proceedings, vol. 972, pp. 69–80. CEUR-WS.org (2012),

38. Park, H., Park, C., Kang, U.: Pegasusn: A scalable and versatile graph mining system. In: McIlraith, S.A., Weinberger, K.Q. (eds.) Proceedings of the Thirty-Second AAAI Conference on Artificial Intelligence, (AAAI-18), the 30th innovative Applications of Artificial Intelligence (IAAI-18), and the 8th AAAI Symposium on Educational Advances in Artificial Intelligence (EAAI-18), New Orleans, Louisiana, USA, February 2–7, 2018. pp. 8214–8215. AAAI Press (2018)
39. Poelmans, J., Ignatov, D.I., Kuznetsov, S.O., Dedene, G.: Fuzzy and rough formal concept analysis: a survey. Int. J. Gen. Syst. **43**(2), 105–134 (2014)
40. Prediger, S.: Logical scaling in formal concept analysis. In: International Conference on Conceptual Structures. pp. 332–341. Springer (1997)
41. Prediger, S., Stumme, G.: Theory-driven logical scaling. In: International Workshop on Description Logics. vol. 22 (1999)
42. Prediger, S., Wille, R.: The lattice of concept graphs of a relationally scaled context. In: International Conference on Conceptual Structures. pp. 401–414. Springer (1999)
43. Raj, P., Raman, A., Nagaraj, D., Duggirala, S.: High-Performance Big-Data Analytics - Computing Systems and Approaches. Computer Communications and Networks, Springer (2015)
44. Rouane, M.H., Huchard, M., Napoli, A., Valtchev, P.: A proposal for combining formal concept analysis and description logics for mining relational data. In: International Conference on Formal Concept Analysis. pp. 51–65. Springer (2007)
45. Scheich, P., Skorsky, M., Vogt, F., Wachter, C., Wille, R.: Conceptual data systems. In: Information and classification, pp. 72–84. Springer (1993)
46. Stumme, G.: Conceptual on-line analytical processing. In: Information Organization and Databases, pp. 191–203. Springer (2000)
47. Stumme, G., Taouil, R., Bastide, Y., Pasquier, N., Lakhal, L.: Computing iceberg concept lattices with titanic. Data & knowledge engineering **42**(2), 189–222 (2002)
48. Sun, Y., Han, J.: Mining Heterogeneous Information Networks: Principles and Methodologies. Synthesis Lect. on Data Mining and Knowledge Discovery, Morgan & Claypool Pub. (2012)
49. Szathmary, L., Valtchev, P., Napoli, A., Godin, R., Boc, A., Makarenkov, V.: A fast compound algorithm for mining generators, closed itemsets, and computing links between equivalence classes. Ann. Math. Artif. Intell. **70**(1–2), 81–105 (2014)
50. Torim, A., Lindroos, K.: Sorting concepts by priority using the theory of monotone systems. In: International Conference on Conceptual Structures. pp. 175–188. Springer (2008)
51. Ventos, V., Soldano, H.: Alpha galois lattices: An overview. In: International Conference on Formal Concept Analysis. pp. 299–314. Springer (2005)
52. Voutsadakis, G.: Polyadic concept analysis. Order **19**(3), 295–304 (2002)
53. Wille, R.: The basic theorem of triadic concept analysis. Order **12**(2), 149–158 (1995)
54. Wu, J., Pan, S., Zhou, C., Li, G., He, W., Zhang, C.: Advances in processing, mining, and learning complex data: From foundations to real-world applications. Complex. **2018**, 1–3 (2018)
55. Wu, X., Zhu, X., Wu, G., Ding, W.: Data mining with big data. IEEE Trans. Knowl. Data Eng. **26**(1), 97–107 (2014)
56. Zadeh, L.A.: Fuzzy sets. Inf. Control. **8**(3), 338–353 (1965). https://doi.org/10.1016/S0019-9958(65)90241-X
57. Zaki, M.J., Hsiao, C.J.: Efficient algorithms for mining closed itemsets and their lattice structure. IEEE transactions on knowledge and data engineering **17**(4), 462–478 (2005)

Chapter 2
Conceptual Navigation in Large Knowledge Graphs

Sébastien Ferré

2.1 Introduction

A growing part of Big Data is made of knowledge graphs. The World Wide Web Consortium[1] has defined a number of standards for representing them (RDF), reasoning about them (RDFS and OWL), and querying them (SPARQL) [18]. A knowledge graph is a collection of entities and values interlinked with semantic relationships. In RDF, every link is a *triple* (source entity, relation, target entity). Examples of entities are *France* and *Paris*, and examples of triples are *(France, capital, Paris)* and *(France, population, 66000000)*. The notion of knowledge graph has been popularized by the Google Knowledge Graph, and the several web search engines have agreed on the `schema.org` vocabulary to support semantic search [25], i.e. search in terms of entities and relationships rather than search in terms of pages and keywords. The two main open sources of knowledge graphs are Linked Open Data[2] (LOD) and microdata.[3] Both count more than 30 billions of triples each. Examples of large open datasets are DBpedia, Wikidata, or YAGO, each of which contains billions of triples.

A major challenge with knowledge graphs is to enable their exploration and querying by people who are interested in the data but are not necessarily proficient in the semantic technologies. A simple kind of exploration is to *browse* the knowledge graph, surfing from entity to entity by following links. However, this kind of exploration can only answer the simplest questions, like *"What is the capital of Paris?"* or *"What is the birth date of the president of the commission of the European Union?"*. We are interested in answering questions that involve sets of entities, and

This research is supported by ANR project PEGASE (ANR-16-CE23-0011-08).

[1] http://www.w3.org/.

[2] See https://lod-cloud.net/.

[3] See http://webdatacommons.org/.

S. Ferré (✉)
Univ Rennes, CNRS, IRISA, Rennes, France
e-mail: ferre@irisa.fr

shared properties, like *"Who are the actors playing most often in films directed by Tim Burton since 2000?"*. The SPARQL query language [31] allows to answer a wide range of questions (high expressivity) but it is a formal language targeted at computer scientists. Even for SPARQL practitioners, writing SPARQL queries is a tedious task because of the need to know the data schema, and because of inevitable trial and errors. Question Answering (QA) approaches [19] are attractive because they rely on spontaneous natural language but they lack expressivity and reliability in practice, because of the challenge of natural language understanding. There exist yet other approaches to help with the exploration and querying of knowledge graphs (e.g., semantic faceted search [2, 17], query builders [20]) but they all tend to trade expressivity for usability.

Formal Concept Analysis (FCA) [14] automatically defines from data a navigational structure, called *concept lattice*, which has been used early for data exploration and analysis [4, 12, 15]. Each concept can be understood as the equivalence class of questions that have the same answers. The *extent* of the concept is that set of answers, and the *intent* of the concept is the most specific question in the equivalence class. The intent represents what all answers have in common. The lattice structure partially orders the concepts according to inclusion between concept extents, with more general concepts (larger extents, smaller intents) at the top, and more specific concepts (smaller extents, larger intents) at the bottom. The concept lattice can be used for exploration and querying as follows. The user starts at the top concept, i.e. with the empty question, and with all entities as answers. She can then move in the lattice down and up according to the lattice structure. For example, she can move down to the concept of *films*, then down again to the concept of *films directed by Tim Burton*, and again to *films directed by Tim Burton since 2000*. From there, she can move down for each actor playing in those films; or she can move up to *films since 2000*. The major advantage of concept lattices is that they materialize a possibly infinite set of questions into a finite navigational structure that is automatically derived from data. However, their major drawback is that their size explodes with the amount of data, and the range of questions.

In this paper, we show how to reconcile expressivity, usability, and scalability in the exploration and querying of knowledge graphs. Compared to a previous paper [7], our approach has improved a lot in expressivity (from a subset of SPARQL 1.0 to most of SPARQL 1.1), in usability (introducing a natural language interface), and in scalability (from tens of thousands of triples to billions of triples). First, to address expressivity, we present an extension of FCA to knowledge graphs, Graph-FCA, where questions and concept intents are analogous to conjunctive SPARQL queries (Sect. 2.2). Second, we introduce *Abstract Conceptual Navigation* (ACN) as an exploration and querying paradigm that reconciles expressivity and usability by relying on the concept lattice to guide end-users in the building of complex queries, and we instantiate it to Graph-FCA (Sect. 2.3). Third, we address scalability on large knowledge graphs by leveraging the computation power of SPARQL endpoints (Sect. 2.4). From there, we demonstrate how our paradigm can rise in expressivity (e.g., logical operators, expressions, aggregations) in an incremental way without losing on usability and scalability (Sect. 2.5). We then present Sparklis, a concrete implementation of our approach available online, and we illustrate it on a few application cases (Sect. 2.6). Readers more interested by practical aspects can read this section first, before diving into the more technical sections. Finally, we conclude and draw a few perspectives (Sect. 2.7).

2.2 Graph-FCA: Extending FCA to Knowledge Graphs

Graph-FCA [8, 11] is an extension of FCA for multi-relational data, and in particular for knowledge graphs of the Semantic Web [18]. The specific nature of Graph-FCA is to extract n-ary concepts from a knowledge graph using k-ary relations. The extents of n-ary concepts are sets of n-ary tuples of graph nodes, and their intents are expressed as a graph pattern with n distinguished nodes, which are called the *projected nodes*. For instance, in a knowledge graph that represents family members with a "parent" binary relation, the "sibling" binary concept can be discovered, and described as *"a pair of persons having a parent in common"*.

Classical FCA corresponds to the case where all relations are *unary* ($k = 1$), i.e. when graph nodes are disconnected, and where all concepts are unary too ($n = 1$), i.e. when concept extents are sets of graph nodes. Graph-FCA differs from Graal [24] and applications of Pattern Structures to graphs [22] in that here objects are the nodes of one large knowledge graph, instead of having each object being described by a small graph. Graph-FCA shares theoretical foundations with the work of Kötters [21], where PGPs are called windowed structures. Another extension of FCA to multi-relational data is Relational Concept Analysis (RCA) [28]. Compared to Graph-FCA, RCA is limited to unary and binary relations and to unary concepts, and its concept intents are tree-shaped graph patterns. However, RCA supports various quantifiers (called *relational scaling operators*, while Graph-FCA is limited to existentials.

2.2.1 Graph Context

Whereas FCA defines a formal context as an incidence relation between objects and attributes, Graph-FCA defines a *graph context* as an incidence relation between *tuples of* objects and attributes. A *graph context* is a triple $K = (O, A, I)$, where O is a set of *objects*, A is a set of *attributes*, and $I \subseteq O^* \times A$ is an *incidence relation* between object tuples $\bar{o} = (o_1, \ldots, o_k) \in O^k$, for any arity k, and attributes $a \in A$. $O^* = \bigcup_{k \in \mathbb{N}} O^k = O \cup (O \times O) \cup (O \times O \times O) \cup \ldots$ denotes the set of all tuples of objects. A graph context is therefore a labeled multi-hyper-graph, where objects are the nodes, incidence elements are the hyper-edges, and attributes are hyper-edge labels. Note that attributes can be interpreted as k-ary predicates, and graph contexts as First Order Logic (FOL) models (without functions and constants). An hyper-edge $((o_1, \ldots, o_k), a)$ can be seen as the FOL *atom* $a(o_1, \ldots, o_k)$, and represents a knowledge *fact*. In the following, we indifferently use the terms *hyper-edge, atom,*

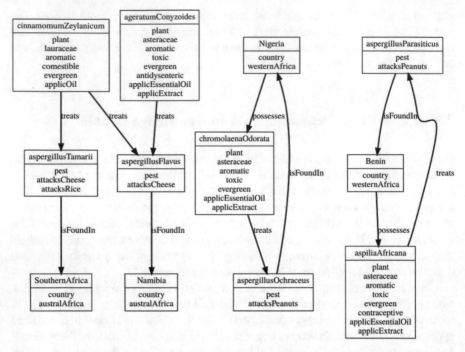

Fig. 2.1: Graph context about plants, pests, and countries. Rectangles are objects, with their labels in the top part, and unary attributes applying to them in the bottom part. Links are binary edges, with the binary attribute as label

and *fact*. Different kinds of knowledge graphs, such as Conceptual Graphs [30], RDF graphs, or RCA contexts, can be directly mapped to a graph context.

We illustrate Graph-FCA on an example taken from project KNOMANA[4] and used in the RCA literature. It is about plants that treat pests found in African countries, which possess such plants. Figure 2.1 shows a graphical representation of the graph context. The objects are plants (e.g. *aspiliaAfricana*), pests (e.g. *aspergillusParasiticus*), and countries (i.e. *Benin*). They are represented as rectangles. The attributes are either unary attributes, like classical FCA attributes (e.g., *plant*, *westernAfrica*, *aromatic*), or binary attributes (e.g., *treats*, *isFoundIn*, *possesses*). The former are represented in the bottom part of object rectangles, and the latter are represented as directed edges between objects. More generally, a unary edge $a(x)$ is represented as a label in the bottom part of the rectangle representing object x; a binary edge $a(x,y)$ is represented by an edge from x to y labeled by a; and other edges $a(x_1,\ldots,x_k)$, for $k > 2$, are represented as ellipses labeled by a, having an edge labeled i to each node x_i.

---

[4] https://ur-aida.cirad.fr/nos-recherches/projets-et-expertises/knomana.

2.2.2 Graph Patterns

Whereas FCA is about finding closed sets of attributes, Graph-FCA is about finding closed graph patterns. A *graph pattern* is similar to a graph context but with variables instead of objects as nodes, in order to generalize over particular objects. A key aspect of Graph-FCA is that closure does not apply directly to graph patterns but to *Projected Graph Patterns* (PGP), i.e. graph patterns with one or several distinguished nodes. Those projected nodes define a projection on the occurences of the pattern, like a projection in relational algebra. For example, the PGP $(x,y) \leftarrow parent(x,z), parent(y,z)$ defines a graph pattern with two edges, $parent(x,z)$ and $parent(y,z)$, and with projection on variables x and y. It means that for every occurence of the pattern in the context, the valuation of (x,y) is an occurence of the PGP. It can be used as a definition of the "sibling" relationship, i.e. the fact that x and y are siblings if they have a common parent z.

PGPs are analogous to anonymous definitions of FOL predicates and to conjunctive SPARQL queries. They play the same role as sets of attributes in FCA, i.e. as concept intents. Set operations are extended from sets of attributes to PGPs. PGP inclusion $\subseteq_q$ is based on graph homomorphisms [16]. It is similar to the notion of *subsumption* on queries [5] or rules [26]. PGP intersection $\cap_q$ is defined as a form of graph alignment, where each pair of variables from the two patterns becomes a variable of the intersection pattern. It corresponds to the *categorical product* of graphs (see [16], p. 116), and to the least general generalization of Plotkin [27].

2.2.3 Graph Concepts

The Galois connection that is the basis for computing concepts is defined in Graph-FCA between n-ary PGPs $(\mathcal{Q}_n, \subseteq_q)$ and n-ary *object relations* $(\mathcal{R}_n, \subseteq)$, where an n-ary object relation is a set of n-ary object tuples $(\mathcal{R}_n = 2^{O^n})$. The connection from PGP $Q \in \mathcal{Q}_n$ to object relation $Q' \in \mathcal{R}_n$ is analogous to query evaluation, and the connection from object relation $R \in \mathcal{R}_n$ to PGP $R' \in \mathcal{Q}_n$ is analogous to relational learning [26]. In the definitions of Q' and R' below, the PGP $(\bar{o}, I)$ represents the description of an object tuple $\bar{o}$ by the whole incidence relation I seen from the relative position of $\bar{o}$.

$$Q' := \{\bar{o} \in O^n \mid Q \subseteq_q (\bar{o}, I)\}, \text{ for } Q \in \mathcal{Q}_n, \text{ for } n \in \mathbb{N}$$
$$R' := \bigcap_q \{(\bar{o}, I)\}_{\bar{o} \in R}, \qquad \text{ for } R \in \mathcal{R}_n, \text{ for } n \in \mathbb{N}$$

From there, concepts can be defined in the usual way, and proved to be organized into lattices. A concept is a pair (Q, R) such that $Q' = R$ and $R' =_q Q$. The arity of the projected tuple of Q must be the same as the arity of object tuples in R. It determines the arity of the concept. Unary concepts are about sets of objects, while binary concepts are about relationships between objects, and so on. Note however

that the intent of an unary concept can mix attributes of different arities. Unlike
RCA, there is a concept lattice for each concept arity rather than for each object type.

Figure 2.2 displays a compact representation of the graph concepts about plants,
pests, and countries as a set of graph patterns. Each rectangle node x identifies a unary
concept (e.g., Q1a) along with its extent (here, *aspiliaAfricana, chromolaenaOdor-
ata*). The concept intent is the PGP $x \leftarrow P$, where P is the subgraph containing node x
and all dark-colored nodes (called the *pattern core*). In the first pattern, all concepts
belong to the core, while on the second pattern, only Q2a, Q2b, Q2c belong to the
core. By reading the graph, we learn that Concept Q1a is the concept of *"plants that
treat pests attacking peanuts, which are found in Western Africa countries that possess the plant, where the plant has a number of features such as being toxic, aromatic,
applicable to essential oil, etc."* This concept therefore identifies the valuable situation where a country possesses plants to treat some pests they have. Concepts Q1b
and Q1c have the same graph pattern as Q1a but a different projected node, on

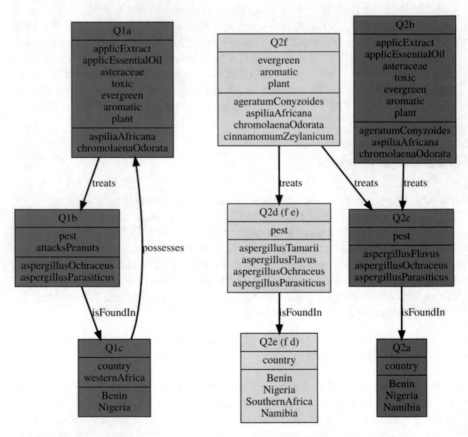

Fig. 2.2: Compact representation of graph concepts about plants, pests, and countries,
with minimum support 2

pests for Q1b and on countries for Q1c. N-ary concepts are obtained by picking several projected nodes. For example, (Q1a,Q1b,Q1c) is a ternary concept whose instances are the object triples (*aspiliaAfricana, aspergillusParasiticus, Benin*) and (*chromolaenaOdarata, aspergillusOchraceus, Nigeria*). It represents the cyclic relationship existing between plants, pests, and countries in Western Africa countries. Pattern Q2 shows that every plant (Q2f) treats some pest (Q2c), and every pest (Q2d) is found in some country (Q2e) but not all countries possess a plant treating a pest. The concept identifiers between bracket, e.g. Q2d(fe), indicate which nodes of the pattern belong to the concept intent, in addition to the core. For Q2d(fe), it means that *"all four pests are found in some country (Q2e), and are treated by some plant (Q2f) that also treats a pest (Q2c) that can be treated by a more specific kind of plant (Q2b), one that is toxic and can be used for essential oils and extracts."*

2.2.4 Graph Concept Lattice

The generalization ordering between concepts, hence (part of) the concept lattice, is shown in Fig. 2.3. The relation from concepts to their children is represented by dashed arrows. The top concept that contains all objects is omitted for clarity. In this representation, the hyper-edges with arity at least 2 are shown inside rectangles with unary attributes. For example in Q1a, the label possesses c _ says that there is a binary edge labeled *possesses* from concept Q1c to this concept Q1a; and label treats _ b says that there is a binary edge from Q1a to Q1b (edges are never across patterns). For example, concepts Q1a, Q1b, Q1c are respectively specializations, hence sub-concepts, of concepts Q2b, Q2c, Q2a. The latter are more general concepts respectively for plants, pests, and countries.

2.3 Conceptual Navigation in Graph-FCA Lattices

The idea of *conceptual navigation* is to see the concept lattice as a navigation structure, with concepts as navigation places, and the lattice structure as a set of navigation links between concepts. In a basic setting, the concept lattice is displayed graphically, and the user can visually navigate it. However, this only works for very small datasets generating at most a few dozen concepts. A more scalable approach consists in displaying only a local view centered on the current concept in the navigation process [1, 6, 13]. Typically, the local view consists of the extent and intent of the current concept, and navigation links to the neighbour concepts. The neighbour concepts are generally the parents and children of the current concept in the Hasse diagram of the concept lattice, but nothing prevents to have links to more distant concepts. Navigation links are generally labeled by the properties that have to be added or removed from the current intent in order to get the intent of the target concept.

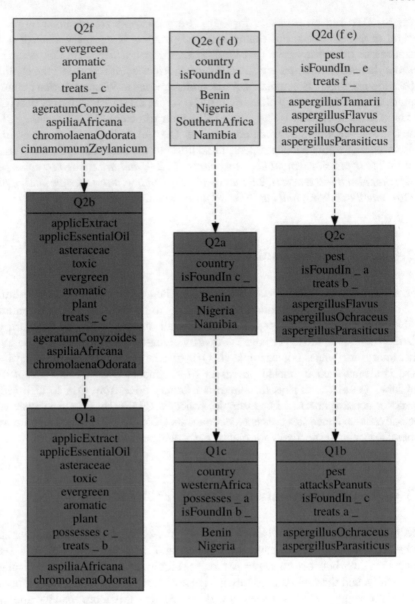

Fig. 2.3: Generalization ordering of unary graph concepts about plants, pests, and countries, with minimum support 2

The major advantage of local views over concept lattices is that the amount of information to be displayed does not grow exponentially with the size of data, like concept lattices, but rather linearly. This fosters expressivity because higher expressivity leads to many more concepts but not to bigger representations of individual concepts. Conceptual navigation with local views also satisfies usability because the user experience is similar to that of faceted search [29], which is commonly used in e-commerce websites. Indeed, users simply have to make a choice among navigation links at each step, and they get a clear view of their current state at all time.

2.3.1 Abstract Conceptual Navigation (ACN)

We first introduce and formalize a very generic framework for conceptual navigation based on local views, called *Abstract Conceptual Navigation* (ACN). We call it abstract by analogy with an abstract class in object-oriented programming, where class members are given a specification but not yet an implementation. The genericity concerns two aspects: (1) the nature of concept intents and extents in order to foster expressivity, and (2) the contents of local views in order to foster usability. In the next section, we instantiate ACN to Graph-FCA concept lattices, where intents are PGPs (projected graph patterns), and extents are object relations. An instantiation on classical FCA would have sets of attributes as intents, and sets of objects as extents.

The main novelty of ACN w.r.t. conceptual navigation, as presented above, is the replacement of the concept intent by a *query* and an *index* in the local view (see Fig. 2.4). The motivation is that concept intents are often at the same time overly specific, and not informative enough. This is because the concept intent is defined as the set of *all and only the* properties that are shared by all elements of the concept extent. The *query* is a subset of the intent that contains only intensional elements selected by the user during her navigation. The query is related to the intent in that both characterize the same extent. The query is useful to the user to let her know the current state of the navigation in a concise way. The *index* is typically a superset of the intent by including properties that are shared only by a subset of the extent, along with their frequencies in the extent. This provides the user with the distribution of

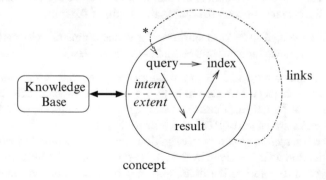

Fig. 2.4: Schema showing the different components of Abstract Conceptual Navigation (ACN), and their interactions

the extent elements over the set of properties: e.g. *"80% films by Tim Burton have genre Fantastic"*. Those properties can then be used as navigation links to more specific concepts. Finally, the extensional component is replaced by *result* to allow for diverse representations of the extent, e.g. tables, maps or charts.

ACN is formally defined by the following components:

knowledge base (*K*). *K* is a *knowledge base* that contains the formal representation of facts, rules, ontological axioms, taxonomies, etc. Its content is constrained by the needs of the application, and the availability of data and domain knowledge. It is an implicit parameter of all ACN operations defined below.

query language (*Q*) and **initial query** (q_0). *Q* is a *query language*, i.e. a set of expressible queries. A *query* is a summary of the navigation history, and expresses the user information needs. The current query $q \in Q$ characterizes the current concept, and is part of its intent. Here, query should be understood in a broad sense, and may include closed and open questions, analytical questions (OLAP), sets of keywords, folder paths, updates, commands, ontological assertions, etc. The initial query q_0 is a fixed query, generally very simple, that determines the starting point of the navigation process.

result (*R* and *result* $\in Q \to R$). *R* is the set of all possible *results* (a.k.a., *query results*), and *result* is a function from queries to their result, given a knowledge base *K*. The query result is any piece of data representing the concept extent, returned to users from the query evaluation. It can be a set of values, an answer list, a table, etc. For some queries, e.g. updates, the knowledge base may be modified as a side effect.

index (*S* and *index* $\in Q \times R \to S$). *S* is the set of all possible indices over the extent, and *index* is a function from queries and their result to their index. An index is at the same time a *summary* over the extent, and a set of query refinements. The intent of the current concept is typically part of the index. Its role is to provide feedback and guidance.

links (*links* $\in Q \times R \times S \to 2^Q$). Function *links* defines a set of navigation links from the current concept to related concepts. Links can be derived from any component of the current concept: the query, the result, or the index.

The main benefit of ACN is to subsume three paradigms of information access: query languages, navigation structures, and interactive views.

- *ACN as a query language. Q* is the query language, and *result* defines query results for each query. Indices and links are void. For instance, SPARQL editors can be seen as partial ACN instances where the query language is SPARQL, and where results are tables of RDF nodes or RDF graphs.
- *ACN as a navigation structure.* Navigation places are specified by queries in *Q*, and navigation links are given by the ACN component *links*. Results and indices may be used to compute the links, and may be used to define the contents of navigation places. For instance, file system hierarchies can be seen as partial ACN instances where queries are directory paths, results are file lists, and navigation links lead to children directories and the parent directory.

- *ACN as an interactive view.* Interactive views follow the MVC architecture (model-view-controller). In ACN, the model is the knowledge base K. The view is the composition of a query, its result, and its index (an element of $Q \times R \times S$). The controller is made of the links, which are derived from the view contents. Each link activation generates a new view by computing a new result (*result*) and a new index (*index*) from the target query. For instance, faceted search can be seen as an ACN instance where a query is a set of facet-values, a result is a selection of items, an index gives the frequency of facet-values among those items, and links allow the addition and removal of facet-values to the query.

ACN inherits *expressivity* from query languages, and *guidance* from navigation structures and interactive views, which achieves our main objective to reconcile *expressivity* and *usability*. However, it must be noted that the actual *expressivity* may be less than the *expressivity* of the query language in the case where navigation links are not rich enough to reach all valid queries. Ideally, links should be both *safe* and *complete*. *Safeness* means that no navigation path leads to dead-ends (e.g., empty results). *Completeness* means that every safe query can be reached through a finite navigation path. *Safeness* is important for the quality of *guidance* as it avoids users to "bump into the walls", and *completeness* is important to leverage the *expressivity* of the query language. Therefore, a critical issue in ACN is the definition of links, and hence the definition of results and indices because links are derived from them.

2.3.2 Graph-ACN: Instantiating ACN to Knowledge Graphs

We here instantiate ACN to Knowledge Graphs, which we call Graph-ACN, by defining each ACN component in Graph-FCA terms (see Sect. 2.2).

Knowledge Base

A knowledge base is formalized as a graph context $K = (O, A, I)$, where objects O are the KG entities, attributes A are the KG k-ary relations, and the incidence relation I is the set of relational facts. As a running example, we consider the graph context displayed in Fig. 2.1.

Query

A query is an unary PGP $Q = x \leftarrow P$, where P is a graph pattern that uses variable x unless it is empty, and that has a single connected component (*connected pattern*). The motivation for excluding disconnected patterns is that each connected component can be navigated to independently of others.

The projected variable x is called the *focus* of the query. It determines the contents of the index, and the behaviour of navigation links. Some navigation links defined below allow users to move it on other variables in the pattern.

The initial query is the empty query $q_0 := x_1 \leftarrow \emptyset$. In our running example, we consider the query

$$q := x_2 \leftarrow plant(x_1), toxic(x_1), treats(x_1, x_2),$$

which corresponds to concept Q2c in Fig. 2.3. It selects *"everything treated by a toxic plant."*

Result

A result is a table with variables in headers, and objects in cells. A result is therefore a pair $r = ((x_1, \ldots, x_n), R)$ composed of a tuple of n variables, and an n-ary object relation $R \in \mathscr{R}_n$. A result is said *empty* if its object relation R is the empty set.

The function from queries to results is defined as

$$result(Q := x \leftarrow P) := (vars(P), (vars(P) \leftarrow P)'),$$

where $vars(P)$ is the row of variables occuring in P. This definition has the same effect as a SELECT * in SPARQL, and provides a richer result than the set of objects Q' that only contains objects at the focus. The result of the initial query q_0 is a one-column table that contains the list of all objects in O. The result of our running query is the following table. Here, there is a bijection between x_1-values and x_2-values but this need not be the case.

$$r := result(q) =$$

x_1	x_2
ageratumConyzoides	aspergillusFlavus
chromolaenaOdorata	aspergillusOchraceus
aspiliaAfricana	aspergillusParasiticus

Index

An index is a set that contains two kinds of elements: variables and *attribute positions*. An attribute position is a pair (a, i) where $a \in A$ is an attribute of arity k belonging to the graph context, and $i \in [1, k]$ is an argument position of the attribute. In the above example about plants, attribute *treats* has two positions: position 1 denotes the plants that treat some pest, while position 2 denotes pests that are treated by some plant.

Given a query $q := x \leftarrow P$ and its result $r = result(q)$, the index $s = index(q, r)$ contains all and only the elements that are *relevant* at focus x. Here, "relevant" means that when the element e is inserted into the query q at focus (as defined right below), the new query has a non-empty result. The insertion function $insert(q, e)$ is defined as follows.

- **Inserting a variable** y modifies the query by replacing every occurences of x by y in q. For example, the insertion of x_1 in the running query q, i.e.

$$insert(q, x_1) = x_1 \leftarrow plant(x_1), toxic(x_1), treats(x_1, x_1),$$

generates a new query where occurences of x_2 is replaced by x_1. That query has obviously an empty result as no pest treats a pest.

- **Inserting an attribute position** (a,i) modifies the query by adding atom $a(y_1,\ldots,y_k)$, where variable y_i is focus variable x and every other variable is a fresh variable. For example, the insertion of $(isFoundIn,1)$ in the running query, i.e.

$$insert(q,(isFoundIn,1)) =$$
$$x_2 \leftarrow plant(x_1), toxic(x_1), treats(x_1,x_2), isFoundIn(x_2,x_3),$$

generates a new query with an additional atom. That query has a non-empty result with an additional column x_3.

x_1	x_2	x_3
ageratumConyzoides	aspergillusFlavus	Namibia
chromolaenaOdorata	aspergillusOchraceus	Nigeria
aspiliaAfricana	aspergillusParasiticus	Benin

The insertion of attribute positions allows to make patterns grow, and the insertion of variables allows to form cycles in patterns. From there, the index can be defined as

$$index(q,r) := \{y \mid y \in vars(P), y \neq x, result(insert(q,y)) \neq \emptyset\}$$
$$\cup \ \{(a,i) \mid a \in A, i \in [1,k], result(insert(q,(a,i))) \neq \emptyset\}$$

A frequency value can be associated to each index element e as the number of objects at focus after inserting it in the query.

$$freq(e) := |insert(q,e)'|$$

Frequencies of attribute positions provide a hint on their distribution over objects at focus. The index of our running query is given in the following table listing index elements along with their frequency.

index element	frequency
$(pest,1)$	3
$(isFoundIn,1)$	3
$(treats,2)$	3
$(attacksPeanut,1)$	2

Links

There are two kinds of links: focus moves and insertion of index elements. A focus move simply changes the focus variable in the query. The insertion of an index element applies function $insert(q,e)$ defined above on any index element $e \in s$. The ACN function *links* is therefore defined as follows, where $q = x \leftarrow P$.

$$links(q,r,s) = \{y \leftarrow P \mid y \in vars(P), y \neq x\}$$
$$\cup \ \{insert(q,e) \mid e \in s\}$$

The number of links is therefore bounded by the number of variables in the current query plus the number of attribute positions. In our running examples, there are 5 links: 1 focus move on x_1, leading to concept Q2b; and 4 insertions for the above index elements, 3 staying on Q2c and the last one leading to Q1b.

Safeness

We define a navigation state as safe in Graph-ACN when it has a non-empty result. Indeed, empty results are like dead-ends in the navigation process, and they are frustrating in a user experience. This is why it should be avoided as much as possible. We can prove safeness by showing that the initial navigation place is safe, and that every navigation link preserves safeness. In our instance, the initial place is obviously safe as the result contains all objects, i.e. all entities of the knowledge graph (otherwise, the KG is empty). Then, there are two kinds of navigation links to consider. Focus moves do not change the query pattern, and by definition of results, they do not change the result either. Therefore, if the source place of the link is safe, then so is the target place. Insertions of index elements also preserve safeness by definition of index elements because their insertion into the current query must lead to queries with non-empty results.

Completeness

We consider Graph-ACN as complete if every connected graph pattern P that has a non-empty extension can be reached in a finite sequence of navigation links, starting from the initial place. We can prove completeness by building a navigation path from the initial query $q_0 = x \leftarrow \emptyset$ to the target query $q = x \leftarrow P$. This can be done by induction. If the pattern has 0 atoms, then it is the initial query, and no navigation link is required. Now, assuming that all patterns with n atoms are reachable, let us consider a pattern P_{n+1} with $n + 1$ atoms. It is always possible to remove a pattern atom $a(x_1, \ldots, x_k)$ from P_{n+1}, and still have a connected pattern P_n with n atoms. Moreover, given that P_{n+1} has a non-empty result, so does P_n because of the Galois connection that exists between PGPs and object relations. By induction hypothesis, there is a navigation path to P_n. From there, P_{n+1} can be reached by moving the focus on a variable x_i that already belongs to P_n (it must exist as P_{n+1} is connected); then by inserting the attribute position (a, i); and finally, for every other position j of the new atom where $x_j \in vars(P_n)$, move the focus on that position, and insert variable x_j.

2.4 Scaling to Large RDF Graphs with SPARQL Endpoints

We here address the concrete implementation of Graph-ACN to knowledge graphs, using semantic web technologies [18]. The objective is to run the navigation process on top of SPARQL endpoints. SPARQL endpoints are web services that can answer complex SPARQL queries over large RDF graphs. Compared to a previous work [7], our approach has scaled from tens of thousands of triples to billions of triples (e.g., DBpedia). In the following we first describe the correspondence between Graph-FCA and both RDF and SPARQL. Then, we describe how to compute results and indices of navigation places from the queries, taking into account that SPARQL engines often return only partial results on large knowledge graphs.

2.4.1 From Graph-FCA to RDF and SPARQL

An RDF graph has three kinds of nodes: URIs (entity identifiers), literals (strings and typed values such as numbers and dates), and blank nodes (anonymous nodes). It has one kind of edge, called *triple* (s, p, o), whose components are respectively called *subject* (the source of the edge), *predicate* (the label of the edge), and *object* (the target of the edge). When a URI is used as a predicate, it is called a *property*, and denotes a binary relationship between RDF nodes. When a URI is used as an object with predicate `rdf:type`, it is called a *class*, and denotes a set of RDF nodes. An RDF graph is simply defined as a set of triples.

In order to apply Graph-ACN on RDF graphs, we need to map RDF graphs to graph contexts. This can be done by applying the following rules:

1. each RDF node n becomes a Graph-FCA object o_n;
2. each URI u and literal l also becomes a Graph-FCA unary attribute, in order to allow for the identification of individual entities and values. This is analogous to nominal scaling in FCA;
3. each RDF class c becomes a Graph-FCA unary attribute;
4. each RDF property p becomes a Graph-FCA binary attribute;
5. *triple* is a Graph-FCA ternary attribute, in order to allow for reified triples;
6. for each URI u, fact $u(o_u)$ represents the identity of the object as a URI;
7. for each literal l, fact $l(o_l)$ represents the value of the object as literal;
8. for each triple $(s, \mathtt{rdf:type}, c)$, fact $c(o_s)$ represents the membership of the subject to the class;
9. for each triple (s, p, o), facts $p(o_s, o_o)$ and *triple*(o_s, o_p, o_o) represent respectively the binary relationship between the subjet and the object, and the reified triple.

It can be observed that the arity of attributes is maximum 3. The unary attributes that are derived from URIs and literals are special in that there is single fact that uses them in the graph context. They are *singleton classes*.

Conceptual navigation in Graph-ACN produces PGPs that have to be translated to SPARQL queries in order to use SPARQL endpoints for the computation of results and indices. A SPARQL query has the form SELECT $?x_1$ ...$?x_n$ WHERE { GP }, where x_is are variable names, and GP is a SPARQL graph pattern. SPARQL has a rich language of graph patterns but we here only need *basic graph patterns*, which are concatenations of *triple patterns*; and equality filters FILTER ($?x = n$) between variable x and RDF node n. A triple pattern s p o. is like an RDF triple except that variables can be used as components in addition to RDF nodes. The function σ translating PGPs Q to SPARQL queries can be defined as follows:

1. $\sigma(Q) = $ 'SELECT ?x_1 ...?x_n WHERE { $\sigma(P)$ }'   with $Q = (x_1, \ldots, x_n) \leftarrow P$;
2. $\sigma(P) = $ '$\sigma(atom_1)$ . ... . $\sigma(atom_m)$'   with $P = \{atom_1, \ldots, atom_m\}$;
3. $\sigma(u(x)) = $ 'FILTER (?x = <u>)'   where u is a URI;
4. $\sigma(l(x)) = $ 'FILTER (?x = "l")'   where l is a literal;
5. $\sigma(c(x)) = $ '?x rdf:type <c>'   where c is a class;
6. $\sigma(p(x,y)) = $ '?x <p> ?y'   where p is a property;
7. $\sigma(triple(x,y,z)) = $ '?x ?y ?z'

The generated SPARQL query can be simplified by eliminating the equality filters. Each equality filter FILTER (?x = n) can be eliminated by replacing in the query all occurences of ?x by n.

2.4.2 Computing the Result, Index, and Links

We here detail the computation of the result, index, and links, assuming that the query q of the current place is the PGP $Q := x \leftarrow P$.

Result

The result of the current place is directly obtained by evaluating the SPARQL translation of the PGP $vars(P) \leftarrow P$. Indeed, the evaluation of a SPARQL query amounts to find all homomorphisms from the graph pattern to the knowledge graph, and it is therefore equivalent to the Galois connection from PGPs to object relations. The output of SPARQL query evaluation is a table with projected variables as headers, and one row per found homomorphism, binding variables to RDF nodes. This is exactly the expected structure for Graph-ACN results.

Index

As defined above, the index is composed of variables and attribute positions whose insertion in q produces a new query with a non-empty result. For efficiency reasons, we want to avoid to compute the result, i.e. to evaluate a SPARQL query, for each variable, and for each attribute position. In other words, we want to decide on condition $result(insert(q,e)) \neq \emptyset$ without actually computing the result nor the insertion.

First, we show that the above condition can be decided for variables only by looking at the current result.

Lemma 2.1 *Let $r := result(q) = ((x_1, \ldots, x_n), R)$, let $x_i = x$ for some $i \in [1,n]$, and let $y = x_j$ for some $j \in [1,n], j \neq i$. The insertion of variable y does not lead to an* *empty result iff there is a row in the current result s.t. both variables x and y have the same value.*

$$result(insert(q,y)) \neq \emptyset \iff \exists (o_1,\ldots,o_n) \in R : o_i = o_j$$

The frequency of index element y can be computed as follows.

$$freq(y) = |\{o_i \mid (o_1,\ldots,o_n) \in R, o_i = o_j\}|$$

The variables in the index, and their frequencies can therefore be computed in one pass over the result.

Second, we show that the above condition can be decided for attribute positions only by looking at the adjacent edges of the focus objects, i.e. by looking at facts that contain an object that is the value of the focus variable in some row of the current result.

Lemma 2.2 *Let $r := result(q) = ((x_1,\ldots,x_n),R)$, let $x_i = x$ for some $i \in [1,n]$, and let (a,j) be an attribute position. The insertion of attribute position (a,j) does not lead to an empty result iff there is a row in the current result, and a fact in the graph context s.t. the value of x in the row is the same as the object at position j in the fact.*

$$result(insert(q,(a,j))) \neq \emptyset \iff$$
$$\exists (o_1,\ldots,o_n) \in R : \exists a(w_1,\ldots,w_k) \in K : o_i = w_j$$

The frequency of index element (a,j) can be computed as follows.

$$freq((a,j)) = |\{o_i \mid (o_1,\ldots,o_n) \in R, a(w_1,\ldots,w_k) \in K, o_i = w_j\}|$$

Assuming that the graph context is only available through the SPARQL endpoint, we need to come up with SPARQL queries that will retrieve the attribute positions. In spirit, we would like to evaluate the conjunctive query $(a,j) \leftarrow P \cup \{a(x_1,\ldots,x_{j-1}, x, x_{j+1},\ldots,x_k)\}$ in order to retrieve all valid attribute positions. Unfortunately, attributes and positions are not valid variables in Graph-FCA queries. However, when considering its translation to SPARQL for the different kinds of attributes available in RDF, we obtain valid SPARQL queries as we show below.

For attributes that represent RDF nodes (1 position), the additional query atom translates as FILTER (?x = ?n), and the SPARQL query simplifies to SELECT ?x WHERE { GP }. This is actually a projection on the focus variable of the query that computes the result. Those attribute positions can therefore be obtained simply by reading column x of the current result, without any request to the SPARQL endpoint.

For attributes that represent RDF classes (1 position), the SPARQL query is SELECT ?x ?c WHERE { GP . ?x rdf:type ?c }. It retrieves all types of focus objects. The inclusion of ?x in the select clauses enables to compute frequencies as a post-processing, and also to use the relationships between focus objects and classes in rich user interfaces.

For attributes that represent RDF properties, there are two positions. The SPARQL query for position 1 is SELECT ?x ?p WHERE { GP . ?x ?p [] }, and for position 2 it is SELECT ?x ?p WHERE { *GP* . [] ?p ?x }, where [] denotes an anonymous variable.

Finally, for attribute *triple* that has 3 positions, there are three queries similar to the queries for properties, except that the additional triple patterns are respectively (?x [] []), ([] ?x []), and ([] [] ?x), and only ?x is in the SELECT-clause.

The above SPARQL queries have the drawback that *GP* has to be evaluated 7 times, including the computation of the result. This is a serious issue for complex patterns. Fortunately, several optimizations are possible. First, all above queries only need the values of x in the pattern. The translated pattern *GP* can therefore be replaced by an enumeration of values for x, as already available in column x of the result: VALUES ?x { o_1 ... o_m }, where $\{o_1, \ldots, o_m\}$ is the projection of result r on column x. This kind of SPARQL pattern is very efficiently evaluated. Second, the results for attribute *triple* at positions 1 and 3 can be deduced from the results about properties because the latter are generalizations of the former. For instance, if $(p, 1)$ is a valid attribute position, then so is $(triple, 1)$. The related SPARQL queries can therefore be dropped. Third, the remaining queries can be factorized by using a UNION-pattern, as follows.

```
SELECT ?x ?c ?p1 ?p2
WHERE {
  VALUES ?x { o1 ... om }

  { ?x rdf:type ?c }
  UNION
  { ?x ?p1 [] }
  UNION
  { [] ?p2 ?x }
  UNION
  { [] ?x [] }
}
```

Each result of that query will bind at most one variable of c, p1, p2 (none of them for attribute position $(triple, 2)$), so that there is no ambiguity in the interpretation of results.

In total, only two requests are done to the SPARQL endpoint per navigation place. The first one to compute the result and part of the index (URIs and literals), and the second one to compute the rest of the index (classes, properties, and *triple*).

Links

The computation of links is immediate once the index is computed. A natural optimization here is to compute the target queries in a lazy way, delaying the actual computation (focus move or index element insertion) to when the user triggers a link.

2.4.3 Living with Partial Results

On large knowledge graphs and for some complex queries, the number of results may be impractical for exhaustive computation, transmission from the endpoint, and display to the end-user. In fact, SPARQL endpoints enforce limits in the number of results (e.g., 10,000 for DBpedia). It is therefore necessary to live with this constraint that entails that only partial results may be available. Using such a limit also allows to tune the responsiveness of the system: the lower the limit, the more responsive the system is.

What are the consequences on Graph-ACN navigation? The fact that the result is partial is not really an issue as it does not make sense to display thousands or millions of results to the end-user. This is similar to web search engines that only show the first 10 results by default, and never show thousands of results at once. Now, a partial result entails a partial index because attribute positions in the index are computed from objects found in the focus column of the result. For RDF nodes (URIs and literals), like for results, it does not make sense to display all of them when there are thousands of them. For classes and properties, their number is generally much smaller than for RDF nodes but in some cases it can still be too high for complete display (e.g., classes in YAGO).

The completeness of conceptual navigation is broken by partial indices. However, experience has shown us that the partial results act as a sample from which the most common classes and properties are retrieved, and the missing index elements are generally unfrequently used. The main limitation of partial indices in practice is when one wants to insert in the query a specific RDF node (e.g., selecting a specific film director among all of them), or occasionally a rare class or property. In order to overcome missing index elements and recover completeness, the solution we have adopted is to let the user enter keywords, and to refine the SPARQL queries with filters using those keywords. For instance, if the user enters keywords "Tim Burton" about focus objects, the SPARQL query computing the result will be SELECT $?x_1 \ldots ?x_n$ WHERE { GP FILTER (REGEX(?x, "Tim", i) && REGEX(?x, "Burton", i)) }. It retrieves only results where the focus value contains words "Tim" and "Burton". When the keywords are about classes and properties, it is the SPARQL query retrieving them that is modified accordingly. A further refinement is to match the keywords on the RDFS labels of URIs rather than on the URIs themselves. We omit the details here.

2.5 Rising in Expressivity

SPARQL has a lot to offer beyond conjunctive queries. It features relational algebra operators (UNION, MINUS, OPTIONAL); scalar expressions that can be used in filters, variable bindings (BIND), and in the SELECT-clause; aggregations with GROUP BY; and yet other features. It is desirable to extend our conceptual navigation framework to cover such expressive features. The difficulty is that operators like

UNION and MINUS do not apply on the whole query, nor on the focus variable but on a subset of query atoms. We therefore need a representation of queries and their focus that is more adequate to the insertion of those SPARQL features.

2.5.1 An Algebraic Form of Queries

We introduce an algebraic form of queries. A *pattern tree* is a pair $T = \langle x, D \rangle$, where x is a variable, and D is a *description* of x possibly containing a number of subtrees. A description D is one of the following forms, where • denotes the described node x:

1. $\top$: the void description;
2. $a(T_1, \ldots, T_{i-1}, \bullet, T_{i+1}, \ldots, T_k)$ with $T_j = \langle x_j, D_j \rangle$ for $j \in [1,k], j \neq i$: a description stating that the described variable is the i-th argument of the hyperedge $a(x_1, \ldots, x_{i-1}, x, x_{i+1}, \ldots, x_k)$, and that each x_j is recursively described by D_j;
3. $\bullet = y$: a description stating that the described variable is equal to y;
4. $D_1 \wedge D_2$: a conjunction of two descriptions, stating that x satisfies both descriptions.

An *algebraic query* is a pattern tree with the focus localized on a part of the tree, either a subtree or a description. Every subtree has a distinct variable, and cycles are expressed with descriptions like $\bullet = y$.

For example, the algebraic query (with focus underlined)

$$\langle x, plant(\bullet) \\ \wedge treats(\bullet, \langle y, isFoundIn(\bullet, \langle z, country(\bullet) \rangle) \rangle) \\ \wedge possesses(\langle \underline{w, \bullet = z} \rangle, \bullet) \rangle$$

is equivalent to the PGP query

$$z \leftarrow plant(x), treats(x,y), isFoundIn(y,z), country(z), possesses(z,x).$$

The algebraic form can be easily obtained from the PGP form $x \leftarrow P$ by using a tree traversal of P, and by putting the focus on the subtree with variable x. For binary attributes, description $a(\bullet, T_2)$ corresponds to cross the binary relation forward, while $a(T_1, \bullet)$ corresponds to cross it backwards.

The computation of the result of algebraic queries requires to translate them to SPARQL, which is defined as follows:

1. $\sigma(T) = \sigma_x(D)$, with $T = \langle x, D \rangle$ (trees)
2. $\sigma_x(\top) = $ "   (void description)
3. $\sigma_x(u(\bullet)) = $ 'FILTER (?x = <u>)'   (URIs)
4. $\sigma_x(l(\bullet)) = $ 'FILTER (?x = "l")'   (literals)
5. $\sigma_x(c(\bullet)) = $ '?x rdf:type <c>'   (classes)
6. $\sigma_x(p(\bullet, \langle y, D_y \rangle)) = $ '?x <p> ?y . $\sigma_y(D_y)$'   (properties)

7. $\sigma_x(p(\langle y, D_y \rangle, \bullet)) = $ '?y <p> ?x . $\sigma_y(D_y)$'   (properties)
8. $\sigma_x(triple(\bullet, \langle y, D_y \rangle, \langle z, D_z \rangle)) = $ '?x ?y ?z . $\sigma_y(D_y)$. $\sigma_z(D_z)$'   (triples)
9. $\sigma_x(triple(\langle y, D_y \rangle, \bullet, \langle z, D_z \rangle)) = $ '?y ?x ?z . $\sigma_y(D_y)$. $\sigma_z(D_z)$'   (triples)
10. $\sigma_x(triple(\langle y, D_y \rangle, \langle z, D_z \rangle, \bullet)) = $ '?y ?z ?x . $\sigma_y(D_y)$. $\sigma_z(D_z)$'   (triples)
11. $\sigma_x(D_1 \wedge D_2) = $ '$\sigma_x(D_1)$. $\sigma_x(D_2)$'   (conjunction)

Finally, it remains to redefine the insertion of variables and attribute positions at the current focus. The insertion of a variable y introduces a new conjunct at the focus. If the focus is on a description D, it is replaced by $D \wedge \bullet = y$. If the focus is on a subtree $T = \langle x, D \rangle$, then the insertion applies on D. Similarly, the insertion of an attribute position (a, i) introduces at the focus the new conjunct $a(T_1, \ldots, T_k)$ where $T_i = \bullet$ and other T_j are fresh variables with void descriptions.

The index and navigation links are then computed in exactly the same way because they rely on the result, which is left unmodified.

2.5.2 Extensions of the Query Algebra

Each extension of the query algebra, aimed at covering some SPARQL feature, consists in introducing new algebraic query constructs. For each new construct, one needs to extend the SPARQL translation, and to define new links to introduce the new construct in a query. The index is generally left unmodified because most additional constructs do not contribute to the description of data. As a consequence, those extensions have almost no impact on usability and scalability. From the user point of view, there are only a few additional navigation links, which can be ignored if she is not interested in the feature. From the point of view of scalability, the query computing the index is the same, and the only impact is on the computation of results depending on which features are used. For example, aggregations are typically costly to evaluate.

We illustrate such an extension with SPARQL UNION, which expresses a disjunction between two patterns. First, we introduce a new description $D_1 \vee D_2$, a disjunction of descriptions. Second, we define its translation to SPARQL, using the UNION operator:

$$\sigma_x(D_1 \vee D_2) = \text{'\{ } \sigma_x(D_1) \text{ \} UNION \{ } \sigma_x(D_2) \text{ \}'}.$$

Third, we add a link from a query whose focus is on a description D to the same query where that description is replaced by $D \vee \top$, introducing the disjunction with an alternative description that is initially void. The focus can be moved on the void description, ready to be refined by insertion of an attribute position for instance.

The query algebra can be extended in a similar way with SPARQL MINUS, which expresses negation on a pattern. The new description is $D_1 \wedge \neg D_2$, whose translation to SPARQL is '$\sigma_x(D_1)$ MINUS { $\sigma_x(D_2)$ }'. The additional link replaces description D at focus by $D \wedge \neg \top$.

Other description extensions express numeric inequalities (e.g., ≥ 10) or string matching (e.g. *contains* "foo", *matches* "[0-9]+"). Query trees can be extended with binding of variables by complex expressions (e.g., computing people's age as the difference between current date and their birthdate), filtering of results by complex expressions (e.g., filtering people whose age is at least 18), and by aggregations over results (e.g., computing the average age of people per country). Expressions constitute a third kind of part in queries, in addition to trees and descriptions, and have thus their own SPARQL translations and navigation links (see [10] for technical details).

An advantage of the algebraic form of queries, beyond rising in expressivity, is that it is more amenable to verbalization in natural language. Indeed, every tree can be verbalized as a noun phrase, while every description can be verbalized as a relative clause. For instance, the example query above can be verbalized as *"a plant that treats something found in a country, and that the country possesses."* This idea has been formalized with the N<A>F design pattern [9].

2.6 The Sparklis Tool and Application Cases

The Graph-ACN framework presented above is implemented in full in the Sparklis tool,[5] though historically Graph-FCA was formalized later than the first implementation. After presenting the implementation, user interface and capabilities of Sparklis, we discuss a few application cases on large knowledge graphs.

2.6.1 Sparklis

Sparklis implements Graph-ACN as a web application that runs on top of SPARQL endpoints. It runs in the web browser as JavaScript (compiled from OCaml with tool js_of_ocaml[6]), and computes extensions and indices by sending SPARQL requests to the SPARQL endpoint.

The user interface of Sparklis is made of a view of the current navigation place, plus configuration widgets. Figure 2.5 shows a screenshot of Sparklis on a core subset of DBpedia. The main configuration widget is a text input at the top where the URL of the SPARQL endpoint has to be entered. Other configuration widgets are available through the "Configure" menu but defaults are generally fine.

The current place view is made of three parts, in accordance with the ACN framework: the query (top), the index (middle), and the result (bottom). The query is fully verbalized in (controlled) natural language, using RDF classes and properties as nouns, and entities as proper nouns. The focus part is highlighted in light green,

---

[5] Available online at http://www.irisa.fr/LIS/ferre/sparklis/.
[6] https://ocsigen.org/js_of_ocaml/3.1.0/manual/overview.

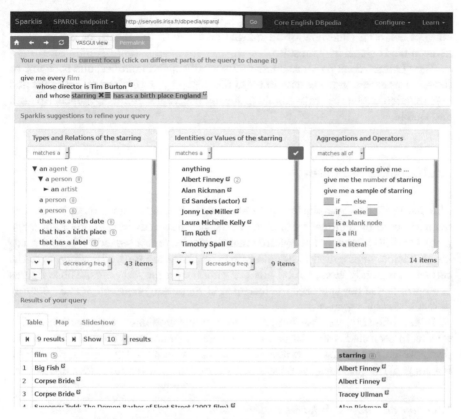

Fig. 2.5: Screenshot of Sparklis where the current query is "the films directed by Tim Burton and starring somebody born in England", with focus on the actors

and can be moved simply by clicking on different parts of the query. At the end of the focus part, the cross allows to delete the focus part, and the hamburger-icon provides a few navigation links to insert operators like conjunction, disjunction, negation, or sorting of results. In the current version, Sparklis covers all SELECT-queries with the exclusion of queries that need disconnected graph patterns.

The index is splitted into three lists: attribute positions for RDF classes, properties, and *triple* (left), attribute positions for RDF nodes (center), and other navigation links for introducing various operators (right). Each index element can be inserted at the query focus simply by clicking it. By default, index elements are sorted by decreasing frequency but they can also be sorted in lexicographic order. Filter inputs at the top of each list allows to force the retrieval of index elements matching some constraint (matching some keywords or satisfying numerical inequalities).

The result is displayed as a table where headers are verbalizations of variables, and the focus column is highlighted in light green. Headers can be clicked to move the focus, and the RDF nodes in the cells can be inserted in the query by clicking them. Depending on the result contents, Sparklis can offer additional views on the concept extent: a slideshow of all images occuring in the result, or a map of all geolocalized objects in the result.

The "Learn" menu provides a large number of example queries, and for some of them, a screencast showing how to build them step by step. The user interface and verbalization of queries are available in several languages: English, French, Spanish and Dutch.

2.6.2 Application Cases

The first version of Sparklis was put online in Spring 2014. Since then, it has been regularly used by thousands of unique users over hundreds of different knowledge graphs (SPARQL endpoints) covering many domains (encyclopaedias, bioinformatics, medicine, bibliography, administrations, etc.). At the time of writing, Sparklis earned consistently around 1000 hits per month over the last years. Those figures demonstrate that our approach is successful in supporting exploration and querying of real knowledge graphs.

In the following, we shortly present four application cases on large knowledge graphs. In each case, we describe the knowledge graph, i.e. its contents, its size, its peculiarities, and we list a few questions that can be built and answered in Sparklis. For some of the questions, we provide a short link to its navigation place in Sparklis.

Encyclopedic Graphs: DBpedia and Wikidata

The most commonly used knowledge graphs are those that have encyclopedic contents. For example, DBpedia [23] is a KG version of Wikipedia, where triples are mostly extracted from the infoboxes included in Wikipedia pages. This implies that DBpedia covers all kinds of topics (e.g., people, organizations, creative works, species, places), and for this reason, it stands in the middle of Linked Open Data (LOD). It is one of the biggest open knowledge graph with more than 3 billions of triples. We list a few questions, taken from QALD-4 challenge, along with the Sparklis link.

1. *Which rivers flow into a German lake?* (https://bit.ly/3uiS3f0)
2. *Give me all films produced by Steven Spielberg with a budget of at least $80 million.* (https://bit.ly/3gkCefE)
3. *How many languages are spoken in Colombia?* (https://bit.ly/34vZapu)
4. *Which poet wrote the most books?* (https://bit.ly/3IYScbC)

Another important encyclopedic KG is Wikidata.[7] Unlike DBpedia, it can be directly edited by people (like a wiki) in a structured way. Editors can create new entities (called *items*), and add *statements* about entities. Some of the contents is

---

[7] https://www.wikidata.org/.

automatically imported from existing structured sources but a lot of its contents is also added manually. Wikidata does not follow the RDF model, although an RDF version exists. In particular, it is possible to make statements about statements. For instance, it is possible to say that Barack Obama is a president of the Unites States, and then say that this presidency started in 2008, and ended in 2016. This specificity and others required several adaptations in Sparklis. Here are a few questions that can be answered in Wikidata.

1. *Show me a map of French hospitals.* (http://bit.ly/2CoTeNP)
2. *What are the popular given names among the French people?* (http://bit.ly/33swvfI)
3. *Give me a slideshow of French prime ministers (since 1815) by decreasing total length of service.* (http://bit.ly/2NtNNUp)

Bibliographic Data at Persée

Persée[8] is a French organization that provides free access to more than 600,000 scientific publications, notably in the domain of humanities and social sciences. It maintains a SPARQL endpoint that gives access to their metadata,[9] and they have officially adopted Sparklis as an exploration and querying tool in 2017. Their motivation was to empower their researchers by enabling them to build complex questions in a free way. The dataset contains about 42 millions of triples. Here are a few questions that they give as example on their web site (translated from French).

1. *Co-authors of Pierre Bourdieu, ordered by lastname and firstname*
2. *Authors with the largest number of co-authors*
3. *For each author, and for each co-author, how many articles in common*
4. *For a given author, the journal titles where he/she published, along with the number of articles and the publication dates*

Pharmacovigilance Data in PEGASE

In the context of the PEGASE[10] research project on pharmacovigilance, we have designed a large knowledge graph about (anonymous) patients taking drugs and having adverse drug reactions (ADR) [3]. It includes patient data from FAERS (about 25 millions of triples for three months of records), and terminological data from SNOMED CT and MedDRA (about 3 millions of triples). The first specificity of this dataset is that there are statements about statements like in Wikidata: e.g., there are statements about how patients have taken drugs (how many, how often, etc.). The second specificity is that MedDRA and SNOMED CT are two large hierarchies that required extensions of Sparklis for querying and visualizing them. There is a maintained SPARQL endpoint but it is not publicly available because of license

---

[8] http://www.persee.fr/.

[9] http://data.persee.fr/.

[10] https://anr.fr/Project-ANR-16-CE23-0011.

restrictions on MedDRA and SNOMED CT. Here are a few examples of questions that can be answered on the PEGASE knowledge graphs.

1. *Are there patients who experienced renal failure after taking Cloxacillin? patients being more than 60 years old?*
2. *Which MedDRA terms describe an inflammatory ADR in the brain?*

Government Data

The QALD-6 challenge (Question Answering over Linked Data) introduced a new task (Task 3) on "Statistical question answering over RDF datacubes" [32]. The dataset contains about 4 million transactions on government spendings all over the world, organized into 50 datacubes, all modeled in RDF. The specificity of the dataset is that it contains a lot of numeric data, and that most questions on it consists in computing aggregations such as averages and totals. The KG contains about 16 millions of triples in total. Unfortunately, the SPARQL endpoint is not maintained so that it is no more possible to explore it in Sparklis. Sparklis could build 148 out of the 150 questions of the challenge. We list here a few questions as examples.

1. *How much was spent on public safety by the Town of Cary in 2010?*
2. *Which expenses had the highest total amount of proposed expenditures for the Maldives?*
3. *How many suppliers did the Newcastle city council use for education?*

2.7 Conclusion and Perspectives

We have introduced *Abstract Conceptual Navigation (ACN)*, a general paradigm based on concept lattices to reconcile expressivity, usability, and scalability in the exploration and querying of a knowledge base. Users are guided in a safe and complete way through the concept lattice, and the user navigation state is represented at all time as a readable query, which is updated according to the navigation links chosen by users. Only a local view centered on the current concept is shown, but ACN gives room to enriched representations of the extent (called *result*) and intent (called *index*) that can provide hints about the concept neighborhood.

We have instantiated ACN to knowledge graphs (Graph-ACN) with Graph-FCA as a formal ground. We have shown how to implement it on top of SPARQL endpoints in a scalable way. In Graph-ACN, queries are conjunctive queries to start with, and are then extended to advanced features available in SPARQL. Results are tables, like for SPARQL queries, thus going beyond the usual sets of objects in other FCA approaches. We have shown an effective implementation of Graph-ACN through the Sparklis tool, and effective applications on real and large knowledge graphs.

In terms of scalability, a perspective is to improve efficiency by designing specialized data structures and algorithms in the SPARQL endpoint to optimize the Graph-ACN-specific SPARQL queries. In terms of expressivity, beyond the SELECT-queries of SPARQL, there are also CONSTRUCT-queries that return graphs, and updates. Beyond SPARQL, we can imagine ACN instances for other query languages such as SQL, XQuery or Cypher. Another perspective for increasing expressivity is to enrich the index in order to support knowledge extraction tasks, e.g. with association rules or functional dependencies.

References

1. Alam, M., Buzmakov, A., Napoli, A.: Exploratory knowledge discovery over web of data. Discrete Applied Mathematics **249**, 2–17 (2018). https://doi.org/10.1016/j.dam.2018.03.041
2. Arenas, M., Grau, B., Kharlamov, E., Š. Marciuška, Zheleznyakov, D., Jimenez-Ruiz, E.: SemFacet: Semantic faceted search over YAGO. In: World Wide Web Conf. Companion, pp. 123–126. WWW Steering Committee (2014)
3. Bobed, C., Douze, L., Ferré, S., Marcilly, R.: Sparklis over PEGASE knowledge graph: a new tool for pharmacovigilance. In: A. Waagmeester, et al. (eds.) Int. Conf. Semantic Web Applications and Tools for Life Sciences (SWAT4LS), *CEUR Workshop Proceedings*, vol. 2275 (2018)
4. Carpineto, C., Romano, G.: A lattice conceptual clustering system and its application to browsing retrieval. Machine Learning **24**(2), 95–122 (1996)
5. Chekol, M.W., Euzenat, J., Genevès, P., Layaïda, N.: SPARQL query containment under SHI axioms. In: AAAI Conf. Artificial Intelligence (2012)
6. Ducrou, J., Eklund, P.: An intelligent user interface for browsing and search MPEG-7 images using concept lattices. Int. J. Foundations of Computer Science, World Scientific **19**(2), 359–381 (2008)
7. Ferré, S.: Conceptual navigation in RDF graphs with SPARQL-like queries. In: L. Kwuida, B. Sertkaya (eds.) Int. Conf. Formal Concept Analysis, LNCS 5986, pp. 193–208. Springer (2010)
8. Ferré, S.: A proposal for extending formal concept analysis to knowledge graphs. In: J. Baixeries, C. Sacarea, M. Ojeda-Aciego (eds.) Int. Conf. Formal Concept Analysis (ICFCA), LNCS 9113, pp. 271–286. Springer (2015)
9. Ferré, S.: Bridging the gap between formal languages and natural languages with zippers. In: H. Sack, et al. (eds.) Extended Semantic Web Conf. (ESWC), pp. 269–284. Springer (2016)
10. Ferré, S.: A SPARQL 1.1 query builder for the data analytics of vanilla RDF graphs. Research report, IRISA, team SemLIS (2018). URL https://hal.inria.fr/hal-01820469
11. Ferré, S., Cellier, P.: Graph-FCA: An extension of formal concept analysis to knowledge graphs. Discrete Applied Mathematics **273**, 81–102 (2019). https://doi.org/10.1016/j.dam.2019.03.003. URL http://www.sciencedirect.com/science/article/pii/S0166218X19301532

12. Ferré, S., Ridoux, O.: A file system based on concept analysis. In: Y. Sagiv (ed.) Int. Conf. Rules and Objects in Databases, LNCS 1861, pp. 1033–1047. Springer (2000)
13. Ferré, S., Ridoux, O.: An introduction to logical information systems. Information Processing & Management **40**(3), 383–419 (2004)
14. Ganter, B., Wille, R.: Formal Concept Analysis — Mathematical Foundations. Springer (1999)
15. Godin, R., Missaoui, R., April, A.: Experimental comparison of navigation in a Galois lattice with conventional information retrieval methods. International Journal of Man-Machine Studies **38**(5), 747–767 (1993)
16. Hahn, G., Tardif, C.: Graph homomorphisms: structure and symmetry. In: Graph symmetry, pp. 107–166. Springer (1997)
17. Hildebrand, M., van Ossenbruggen, J., Hardman, L.: /facet: A browser for heterogeneous semantic web repositories. In: I.C. *et al* (ed.) Int. Semantic Web Conf., LNCS 4273, pp. 272–285. Springer (2006)
18. Hitzler, P., Krötzsch, M., Rudolph, S.: Foundations of Semantic Web Technologies. Chapman & Hall/CRC (2009)
19. Höffner, K., Walter, S., Marx, E., Lehmann, J., Ngomo, A.C.N., Usbeck, R.: Overcoming challenges of semantic question answering in the semantic web. Semantic Web Journal (2016)
20. Kaufmann, E., Bernstein, A.: Evaluating the usability of natural language query languages and interfaces to semantic web knowledge bases. J. Web Semantics **8**(4), 377–393 (2010)
21. Kötters, J.: Concept lattices of a relational structure. In: H. Pfeiffer, and others (eds.) Int. Conf. Conceptual Structures for STEM Research and Education, LNAI 7735, pp. 301–310. Springer (2013)
22. Kuznetsov, S.O., Samokhin, M.V.: Learning closed sets of labeled graphs for chemical applications. In: S. Kramer, B. Pfahringer (eds.) Int. Conf. Inductive Logic Programming, LNCS 3625, pp. 190–208. Springer (2005)
23. Lehmann, J., Isele, R., Jakob, M., Jentzsch, A., Kontokostas, D., Mendes, P.N., Hellmann, S., Morsey, M., van Kleef, P., Auer, S., Bizer, C.: DBpedia - a large-scale, multilingual knowledge base extracted from wikipedia. Semantic Web Journal (2013). Under review.
24. Liquiere, M., Sallantin, J.: Structural machine learning with galois lattice and graphs. In: Int. Conf. Machine Learning, pp. 305–313 (1998)
25. Mika, P.: On schema.org and why it matters for the web. IEEE Internet Computing **19**(4), 52–55 (2015)
26. Muggleton, S., Raedt, L.D.: Inductive logic programming: Theory and methods. Journal of Logic Programming **19,20**, 629–679 (1994)
27. Plotkin, G.: Automatic methods of inductive inference. Ph.D. thesis, Edinburgh University (1971)
28. Rouane-Hacene, M., Huchard, M., Napoli, A., Valtchev, P.: Relational concept analysis: mining concept lattices from multi-relational data. Annals of Mathematics and Artificial Intelligence **67**(1), 81–108 (2013)

29. Sacco, G.M., Tzitzikas, Y. (eds.): Dynamic taxonomies and faceted search. The information retrieval series. Springer (2009)
30. Sowa, J.: Conceptual structures. Information processing in man and machine. Addison-Wesley, Reading, US (1984)
31. SPARQL 1.1 query language (2012). URL http://www.w3.org/TR/sparql11-query/. W3C Recommendation
32. Unger, C., Ngomo, A.C.N., Cabrio, E.: 6th open challenge on question answering over linked data (QALD-6). In: H. Sack, et al. (eds.) Semantic Web Evaluation Challenge, pp. 171–177. Springer (2016)

Chapter 3
FCA2VEC: Embedding Techniques for Formal Concept Analysis

Dominik Dürrschnabel, Tom Hanika, and Maximilian Stubbemann

3.1 Introduction

A common approach for the study of complex data sets is to embed them into appropriate sized real-valued vector spaces, e.g., $\mathbb{R}^d$, where d is a small natural number with respect to the dimension of the original data. This enables the application of the well understood and extensive toolset from linear algebra. The practice is propelled by the presumption that relational and other features from the data will be translated to positions and distances in $\mathbb{R}^d$, at least up to some extent. Especially relative distances of embedded entities are often meaningful [19]. For example, Wang et al. employed the addressed embeddings for the complex data structure knowledge graphs (KG) [32]. Particularly Ristoski et al. [22, 26] demonstrated an embedding of the Wikidata KG [31] in a 100-dimensional vector space.

The embedding approaches were shown to be successful for many research fields such as *link prediction*, *clustering*, and *information retrieval*. Hence, today they are widely applied despite that they also do exhibit multiple shortcomings. One of the most pressing is the fact that learned embeddings elude themselves from interpretation [12] and explanation, even though they are conducted in comparatively low dimension, e.g., 100. In our work we want to overcome this disadvantage. We

---

Authors are given in alphabetical order. No priority in authorship is implied.

D. Dürrschnabel • T. Hanika (✉)
Knowledge & Data Engineering Group, University of Kassel, Kassel, Germany
Interdisciplinary Research Center for Information System Design, University of Kassel, Kassel, Germany
e-mail: duerrschnabel@cs.uni-kassel.de; hanika@cs.uni-kassel.de

M. Stubbemann
Knowledge & Data Engineering Group, University of Kassel, Kassel, Germany
Interdisciplinary Research Center for Information System Design, University of Kassel, Kassel, Germany
L3S Research Center, Leibniz University Hannover, Hannover, Germany
e-mail: stubbemann@cs.uni-kassel.de

© The Author(s), under exclusive license to Springer Nature Switzerland AG 2022
R. Missaoui et al. (eds.), *Complex Data Analytics with Formal Concept Analysis*,
https://doi.org/10.1007/978-3-030-93278-7_3

step in with an exploration of the connection of formal concept analysis notions on the one side and proven embedding methods such as *word2vec* or *node2vec* on the other. Our investigation is two-fold and can be represented by two questions: First, how can vector space embeddings be exploited for coping more efficiently with problems from FCA? Secondly, to what extent can conceptual structures from FCA contribute to the embedding of formal context like data structures, e.g., bipartite graphs? In order to deal with the aforementioned shortcomings, namely the non-interpretability/explainability, we limit our search for results to the posed questions using an additional constraint. We want to make use of only two and three dimensions for the real-valued vector space embeddings. By doing so, we ensure the possibility of human comprehension, interpretation or even explanation of the results, at least to some extent.

Equipped with this problem setting we perform different theoretical and practical considerations. We revisit previous work by Rudolph [28] and propose different models for learning closure operators using neural networks. Conversely to this we develop a natural procedure for improving low dimensional node embeddings using formal concepts. We evaluate the introduced techniques by experiments on covering relations and link prediction. Finally, we present some first ideas on how to treat, identify and extract implications using the partially learned closure operators.

The following is divided into five sections. First, we start with an overview over related work in Sect. 3.2. This will include in particular previous work from FCA. In Sect. 3.3 we recollect operations and notations from formal concept analysis and word2vec. The next section contains our modeling which connects the field of FCA with word2vec-like approaches. Here we provide some theoretical insights into what aspects of closure operators can be learned through embeddings. This is followed by an experimental evaluation in Sect. 3.5 employing three medium-sized data sets, i.e., the well-known *Mushroom* context, a dense extract of the Wikidata KG, called *wiki44k* [15], and a bipartite publication graph consisting of authors in the FCA community.[1] We conclude our work with Sect. 3.6 providing further research questions to be investigated.

3.2 Related Work

We will employ in our work learning models based on neural networks, in particular, but not limited to, word2vec [19] and derived works such as node2vec [13]. Multiple approaches that aim to establish ties between FCA and neural networks were proposed in the past. For example, [17] uses concept lattices to derive network architectures for classification tasks while [7] uses a NN to compute concept lattices. However, to the best of our knowledge, there are no previous works on embedding (FCA) closure systems into real-valued vector spaces using a neural network (NN)

---

[1] The data was extracted from https://dblp.uni-trier.de/ and is part of the testing data set for the formal concept analysis software *conexp-clj*, which is hosted at GitHub, see https://github.com/tomhanika/conexp-clj/tree/dev/testing-data.

based learning setting. There is an plethora of principle investigations for embedding *finite ordinal data* in real vector spaces, first of all *measurement structures* [30] (which we found via [34]). In there the author investigated the basic feasibility of such an endeavor. Along this line of research in the realm of FCA is [33], which is focused on *ordinal formal contexts*.

A commonly used approach for the reduction of formal contexts is binary matrix decomposition [4, 5]. However, we are only aware of one FCA-based learning model that computes real valued vector representations of objects and attributes: in [8], the authors transfer latent semantic analysis (LSA) to the realm of FCA. Their analysis demonstrates that the LSA learning procedure does lead to useful structures. Nonetheless, we refrain from considering LSA for our work. The NN procedures we investigate possesss a crucial advantage over LSA: they are able to cope with incremental updates of the relational data efficiently [24]. In the realm of modern complex data structures, such as Wikidata, this is a necessity.

More research in the line of such data structures, namely Resource Description Framework Graphs, was explored by the works [22, 26, 32]. For this the authors use simple as well as sophisticated approaches. The overall goal in these compositions is to provide node similarity corresponding to the underlying relational structure. Since our goal is to excavate and employ a hidden conceptual relation we will develop an alternative NN method for formal context like data. For this we also foster from [28]. In there the author conducts a more fundamental approach for employing NN in the realm of closure systems. Notably, an encoding of closure operators through NN using FCA is presented.

3.3 Foundations

The foundations can be split into two parts: formal concept analysis and word2vec-based embedding approaches. Throughout this work, the term embedding always refers to an general procedure or function, that maps some kind of structure (graphs, nodes, words) to a real-valued vector space. This notion is commonly used in machine learning. It should not be confused with the strict embedding term from mathematics, which denotes a structure-preserving injection.

3.3.1 Formal Concept Analysis

Before we start with our modeling, we want to recall necessary notions from formal concept analysis. For a detailed introduction we refer the reader to [11]. A *formal context* is a triple $\mathbb{K} := (G, M, I)$, where G represents the finite *object set*, M the finite *attribute set*, and $I \subseteq G \times M$ a binary relation called *incidence*. We say for $(g, m) \in I$ that object $g \in G$ has attribute $m \in M$. In this structure we find a (natural) pair of derivation operators $\cdot' : \mathscr{P}(G) \to \mathscr{P}(M)$, $A \mapsto A' := \{m \in M \mid \forall g \in A \colon (g, m) \in I\}$

and $\cdot' : \mathscr{P}(M) \to \mathscr{P}(G)$, $B \mapsto B' := \{g \in G \mid \forall m \in B : (g,m) \in I\}$. Those give rise to the notion of a *formal concept*, i.e., a pair (A,B) consisting of an object set $A \subseteq G$, called *extent*, and an attribute set $B \subseteq M$, called *intent*, such that $A' = B$ and $B' = A$ holds. The set of all formal concepts $(\mathfrak{B}(\mathbb{K}))$ with the order $(A_1, B_1) \leq (A_2, B_2) :\Leftrightarrow A_1 \subseteq A_2$ constitutes a lattice [11], the so called *formal concept lattice*. It is denoted by $\underline{\mathfrak{B}}(\mathbb{K}) := (\mathfrak{B}(\mathbb{K}), \leq)$. Throughout this work we consider formal contexts with $\forall g \in G : \{g\}' \neq \emptyset$ and $\forall m \in M : \{m\}' \neq \emptyset$.

3.3.2 Word2Vec

We adapt the word2vec approach [19, 20] that generates vector embeddings for words from large text corpora. The model gets as input a list of sentences. It is then trained using one of two different approaches: predicting for a target word the context words around it (the *Skip-gram* model, called SG); predicting from a set of context words a target word (the *Continuous Bag of Words* model, called CBOW). In detail, word2vec works as described in the following.

Let $V = \{v_1, \ldots, v_n\}$ be the vocabulary. We identify V as a subset of the vector space $\mathbb{R}^n$ via $\phi : V \to \mathbb{R}^n, v_i \mapsto e^i$, the i-th vector of the standard basis of $\mathbb{R}^n$. This identification is commonly known under the term *one-hot encoding*. The learning task then is the following: Find for a given $d \in \mathbb{N}$ with $d \ll n$ a linear map $\varphi : \mathbb{R}^n \to \mathbb{R}^d$, i.e., a matrix $W \in \mathbb{R}^{d \times n}$ which obeys the goal: words that appear in similar contexts shall be mapped closely by φ. The final embedding vectors of the words of the vocabulary are given by the map

$$\Upsilon : V \to \mathbb{R}^d, v \mapsto \varphi(\phi(v)). \tag{3.1}$$

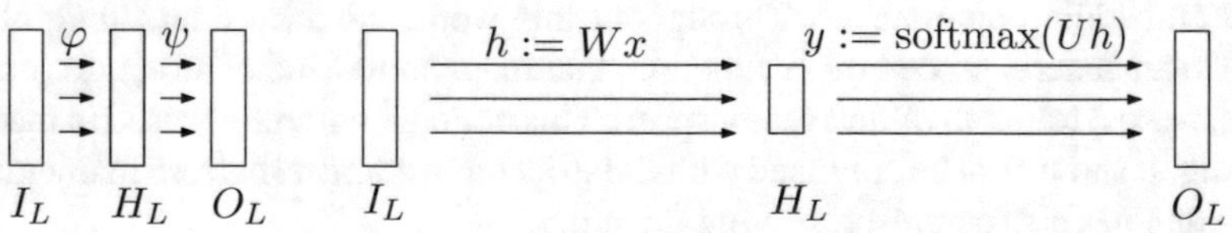

Fig. 3.1: Left: A generic neural network consisting of 3 layers. Right: The structure of the word2vec architecture. The neural network consists of an input and output layer of size n and a hidden layer of size d, where $d \ll n$

To obtain such an embedding, word2vec uses a neural network approach. This network consists of two linear maps and a softmax activation function, cf. Fig. 3.1. The first linear function maps the input from $\mathbb{R}^n$ to $\mathbb{R}^d$, the second one from $\mathbb{R}^d$ back to $\mathbb{R}^n$. In detail, the neural net function has the structure

$$\text{NN} : \mathbb{R}^n \to \mathbb{R}^n, x \mapsto \text{softmax}(\psi(\varphi(x))),$$

where $\varphi : \mathbb{R}^n \to \mathbb{R}^d$ and $\psi : \mathbb{R}^d \to \mathbb{R}^n$ are linear maps with corresponding matrices $W \in \mathbb{R}^{d \times n}$ and $U \in \mathbb{R}^{n \times d}$. The softmax activation function is given by

$$\text{softmax} : \mathbb{R}^n \to \mathbb{R}^n, \begin{pmatrix} x_1 \\ \vdots \\ x_n \end{pmatrix} \mapsto \frac{1}{\sum_{l=1}^n \exp(x_l)} \begin{pmatrix} \exp(x_1) \\ \vdots \\ \exp(x_n) \end{pmatrix}.$$

The function φ is then used in word2vec for creating embeddings of the words $v \in V$ via Eq. (3.1). If we use the notion of layers, as described in [6], the neural network function is a three-layer network, consisting of an input (I_L), hidden (H_L), and output layer (O_L). In this notation the hidden layer is used to determine the embeddings. We refer the reader again to Fig. 3.1. In the realm of word2vec Mikolov et al. [19] proposed two different approaches to obtain the matrices W and U from input data. Those are called the Skip-gram and the Continuous Bag of Words architecture. We recollect them in the following.

3.3.2.1 The Skip-Gram and the Continuous Bag of Words Architecture

The SG architecture trains the network to predict for a given *target word* the *context words* around it. Each training example consist of a target word and a finite sequence of context words. We formalize these as tuples $(t, (c_i)_{i=0}^l) \in V \times V^{<\mathbb{N}}$, where $V^{<\mathbb{N}}$ is the set of finite sequences of elements of V. The SG architecture generates the input-output pairs $(\phi(t), \phi(c_0)), \ldots, (\phi(t), \phi(c_l))$ as training examples, where ϕ is the one-hot encoding function, as introduced above. In the CBOW model, in contrast to SG, the training pairs are generated differently from $(t, (c_i)_{i=0}^l)$. We take the middle point of the vectors $\phi(c_0), \ldots, \phi(c_l)$ and try to predict the target word $\phi(t)$, hence the generated input-output training pair is $(\frac{1}{l} \sum_{i=0}^l \phi(c_i), \phi(t))$. Both architectures employ the same kind of loss function to learn the weights of W and U. The error term is computed through cross-entropy loss. The backpropagation is done via stochastic gradient descent. A detailed explanation can be found in [27].

In word2vec, the pairs of target word and context words are generated from text data sequences, i.e., lists of sentences. The word2vec approach has a window size $m \in \mathbb{N}$ as a parameter, i.e., for a given sentence $s = (w_i)_{i=0}^l \in V^{<\mathbb{N}}$ pairs of target word and context word sequences are defined in the following manner. For every $i \in \{0, \ldots, l\}$ a reduced window size $m_i \in \{0, \ldots, m\}$ is chosen randomly and the pair $(w_i, (w_{i+k})_{k=-m_i, k \neq 0}^{m_i})$ is used as pair of target word and context word sequence. Of course, m_i is to be chosen reasonable with respect to l.

Note, in the case of word embeddings, the size of the vocabulary often reaches a level where computing the softmax is computationally infeasible. Hence, the softmax layer is often approximated/replaced by one of the two following approaches. The *hierachical softmax* [21] stores the elements of the vocabulary in a binary Huffmann tree and then only uses the values to the path to an element to compute its probability. Another approach presented in [20] is *negative sampling*, which uses the sigmoid function. In the experimental part of this work we deal with formal contexts of a size where applying the softmax layer is possible.

3.4 Modeling

This section is split in two parts following the two mentioned research directions. In the first part we demonstrate how embeddings can be used in order to retrieve (classical) FCA relevant features from data. We will discover that different aspects of closure operators can be encoded into real-valued vector space embeddings through neural network techniques. In particular, we are looking at covering relations as well as canonical bases. In the second part we propose a straightforward approach for embedding objects and attributes with respect to their conceptual structure. While the first part deviates from the classical word2vec approach due to theoretical considerations, the second part translates FCA notions to the word2vec model. Both investigations are governed by the overall goal from FCA to create explainable methods. To this end we apply for all our methods only low dimensional embeddings, i.e., two or three dimensions. Hence, these embeddings comprise the potential for human interpretability or even explainability, in contrast to high dimensional ones.

3.4.1 Retrieving FCA Features Through Closure2Vec

The goal of this section is to employ the ideas from word2vec to improve the computational feasibility of common tasks in the realm of formal concept analysis. Doing so, we analyze different approaches and finally settle with a novel embedding technique that can provide more efficient computations. In particular, we consider the FCA problems of finding the covering relation of the concept lattice structure and the rediscovering of canonical bases. The linchpin of our investigation is the encapsulation of the closure operator of a formal context.

Analogue to the approach of word2vec we want to achieve a meaningful embedding of the closure operator into $\mathbb{R}^d$ for $d = 2$ or $d = 3$. For the rest of this part we assume that both the attribute set and the object set are indexed, i.e., for some context (G, M, I) we denote the object set by $G = \{g_1, g_2, \dots\}$ and the attribute set by $M = \{m_1, m_2, \dots\}$. This enables the possibility for defining the *binary encoding* of an object set A as the vector $v \in \{0, 1\}^{|M|}$, with the $v_i = 1$ if and only if $m_i \in A$. Dually this can be done for attribute sets.

3.4.1.1 Exact Representation of the Closure Operator

Thanks to a previous work by S. Rudolph [28], we are aware that it is possible to represent any closure operator on a finite set into a neural network function using formal concept analysis. The network, as proposed in [28], consists of an *input layer* $I_L := \{0,1\}^{|M|}$, a *hidden layer* $H_L := \{0,1\}^{|G|}$ as well as an *output layer* called $O_L := \{0,1\}^{|M|}$. The mapping between I_L and H_L is defined as $\varphi = t \circ w$ consisting of a linear mapping w with transformation matrix $W = (w_{jh}) \in \{-1,0\}^{|M| \times |G|}$, such that

$$w_{jh} := \begin{cases} 0 & \text{if } (g_j, m_h) \in I, \\ -1 & \text{otherwise,} \end{cases}$$

and a non-linear activation function $t : \mathbb{R}^{|G|} \to \mathbb{R}^{|G|}$ with each component being mapped using the function $\tilde{t} : \mathbb{R} \to \{0,1\}$ defined as

$$\tilde{t}(x) = \begin{cases} 1 & x = 0, \\ 0 & x < 0. \end{cases}$$

The mapping between H_L and O_L is defined analogously by $\psi = \hat{t} \circ \hat{w}$, where once again $\hat{w}$ is a linear mapping with transformation matrix $\hat{W} = W^T$. The function $\hat{t} : \mathbb{R}^{|M|} \to \mathbb{R}^{|M|}$ is once again defined component-wise with each component being $\tilde{t}$. Using this construction the function $\varphi \circ \psi$ encapsulates the closure operator. To find the closure of some attribute set $B \subseteq M$, one has to compute $\varphi \circ \psi$ of its binary encoding. Similar to both derivation operators introduced in Sect. 3.3, the mappings φ and ψ compute the attribute- and the object derivation, both in their binary encoding.

	1	2	3
a		×	×
b	×		×
c		×	

Fig. 3.2: A formal context counterexample for Proposition 3.1

3.4.1.2 Considering the Unconstraint Problem

Considering the well established word2vec architecture the following idea seems intuitive. Take the neural network layers as defined in Rudolph's architecture, but replace the hidden layer by a layer containing either two or three dimensions, i.e., we have $H_L = \mathbb{R}^d$. Instead of presetting φ and ψ, as in the last section, we want to retrieve them through machine learning. However, it may be noted that it is not meaningful to allow arbitrary and unconstrained functions. To see this, consider the following example with $d = 1$. Let $s : \{0,1\}^{|M|} \to \mathbb{N}$ be an injective mapping from the set of binary vectors of length $|M|$ to the natural numbers. Naturally there is an inverse map $s^{-1} : s[\{0,1\}^{|M|}] \to \{0,1\}^{|M|}$, where $s[\{0,1\}^{|M|}]$ denotes the image of

s of the domain. Since $\mathbb{N}$ is contained in H_L, we may find a natural continuation of s^{-1} to $\mathbb{R}$ by $\overline{s^{-1}} : \mathbb{R} \to \{0,1\}^{|M|}$ such that $\overline{s^{-1}}(x) := s^{-1}(\lfloor x \rfloor)$. Furthermore, let cl be the double application of the derivation operator, i.e., $(\cdot')'$ in the binary encoding. Using this setup let $\varphi := s$ and $\psi := \overline{s^{-1}} \circ cl$. Using these functions one can easily see that even though the neural network is able to compute the closure operator, the layer H_L contains no information about the formal context. This suggests that the set of possible functions has to be further constrained.

3.4.1.3 Representing Closure Operators Using Linear Functions

Rudolph's approach for representing a closure operator by a neural network function is sound and complete. It consists, as discussed, of two linear functions and two non-linear activation functions. The latter, however, is incompatible with the neural network proposed by word2vec. This procedure, as noted in Sect. 3.3.2, does consist of a linear map φ from I_L to H_L, which is also the final embedding we are looking for in our work. Note that it is not possible to represent a closure operator using a linear function, since the closure of the empty set is not necessarily an empty set. The same fact is true for affine mappings, as showed in the following.

Proposition 3.1 *Let (G,M,I) be a formal context. The set of all affine linear mappings, which represent the closure operator on the attribute set in binary encoding, can be empty.*

Proof Consider the formal context from Fig. 3.2. For the sake of simplicity we speak about attribute and object sets and their respective binary encodings interchangeably. Assume that there is an affine mapping which maps each attribute set to its closure. Then there is a linear mapping l such that for each attribute set $v \in \{0,1\}^{|M|}$ the vector $v' = [1 \ v]$ is mapped to the closure of v. Here $[1 \ v]$ denotes the vector which results from the concatenation of a single bit (valued 1) with v. Using this one can infer that from

$$\{\}'' = \{a,b,c\}' = \{\}$$
$$\{3\}'' = \{a,b\}' = \{3\}$$
$$\{1,2\}'' = \{\}' = \{1,2,3\}$$
$$\{1,2,3\}'' = \{\}' = \{1,2,3\}$$

follows that

$$l(1,0,0,0) = (0,0,0)$$
$$l(1,0,0,1) = (0,0,1)$$
$$l(1,1,1,0) = (1,1,1)$$
$$l(1,1,1,1) = (1,1,1).$$

However, as l is a linear mapping, it is required that the following holds.

$$l(1,1,1,1) = l(1,1,1,0) + l(1,0,0,1) - l(1,0,0,0)$$
$$= (1,1,1) + (0,0,1) - (0,0,0)$$
$$= (1,1,2),$$

Hence, we obtain a contradiction.   □

Corollary 3.1 *Let (G,M,I) be a formal context. The set of all affine linear mappings, which represent the derivation operator on the attribute set in binary encoding, can be empty.*

Proof Assume there is such an affine linear map a. By duality we know that there must be an affine linear map a^d on the object set. A suitable composition of those mapping, i.e., using augmentation, contradicts with Proposition 3.1.   □

3.4.1.4 Linear Representable Part of Closure Operators

We know from the last section that it is neither possible to represent the closure operator nor the derivation operator using an affine linear function. Still, it might be possible to obtain a meaningful approximation of an embedding using a linear map. In order to obtain some empirical evidence if studying this approach is fruitful we conduct a short experiment. Consider the neural network architecture as depicted in Fig. 3.1 (left). Furthermore, let the input layer I_L of size $|M|+1$ be connected to a hidden layer H_L of low dimension, i.e., two or three, by a linear function φ. The layer H_L is connected to the output layer O_L that is of dimension M using a function ψ that consists of a linear function together with a *sigmoid* activation function. The first bit of I_L is always set to 1 and therefore a so-called bias unit. For our experiment we now train the neural network by showing it randomly sampled attribute sets as inputs and their attribute closures as output, both in their binary encoding. We employ for this mean squared error as the loss function and a learning rate of 0.001. Even though the

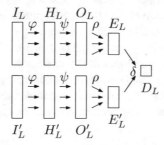

Fig. 3.3: The siamese neural network used to compute *closure2vec*. The functions φ, ψ, ρ are shared functions between the layers in this model

neural network starts to memorize the samples it has seen after around 20 epochs, it does not generalize to attribute sets not previously seen in training. Furthermore, the resulting embedding into $\mathbb{R}^d$ does empirically not expose a meaningful structure. Additionally, this observation does not alter by changing the function ψ to a linear function. Also, experiments in which we investigated learning only the derivative operator were not fruitful. This is the expected behavior from our considerations in the last section. We do not claim that there are no better performing approaches for this task. Still this result motivates our progression to a different task.

3.4.1.5 Non-linear Embedding Through Closure2Vec

As linear embeddings do not seem to work out for our learning task, we employ a different approach. Let the *closure Hamming distance (CHD)* for two attribute sets $A \subseteq M$ and $B \subseteq M$ be the distance function $d(A,B) := d_H((A'')^b, (B'')^b)$, where $\cdot^b$ denotes the binary representation and d_H is the Hamming distance. Note that the closure Hamming distance is not a metric, as the distance between two attribute sets sharing the same closure is 0, even though they are not the same. Based on the idea that two attribute sets are similar if they have a small CHD, we want to embed the attribute sets into a low-dimensional real-valued space, i.e., two or three dimension. The goal here is that the embedding is approximately an isometric map.

We train a neural network architecture that we call *closure2vec* to learn the just introduced CHD. For this, consider the network depicted in Fig. 3.3. It consists of two input layers I_L and I'_L, each of size $|M|$. Then the function φ consisting of a linear function and a *ReLU-activation* function (see [18]) is used to feed the data into the hidden layers H_L and H'_L respectively, both of size $|G|$. After this the function ψ, consisting of a linear function and a ReLU-activation function is applied to the two "streams" in the network. The result then is input for two output layers O_L and O'_L, both of size $|M|$. This, however is not the final step of this network model. Finally, the layers E_L and E'_L, which consist of either two or three dimensions, are fed from O_L and O'_L, respectively, via ρ. This function is again built via composing a linear function and another ReLU-activation function. The output layer D_L has size one. Using a fixed function δ (in our case either the Euclidean distance or cosine distance) we compute a distance between O_L and O'_L. By sharing the functions φ, ψ, and ρ between the different layers we ensure that a commutation of the input sets does not lead to a different prediction of the neural network.

The network then is trained by showing it two attribute sets in binary encoding as well as their closure Hamming distance at the output layer. The required loss function for this setup is then the mean squared error. The learning rate of our network is set to 0.001. The training set for our approach is sampled as follows: For some $t \in \mathbb{N}$ take all attribute combinations that contain at most t elements and put them in some set $\mathscr{X} = \{X_1, X_2, \dots\}$, hence, $X_i \subseteq M$. For each $X_i \in \mathscr{X}$ generate a random attribute $m_i \in M$. Let the set $\mathscr{Y} = \{Y_1, Y_2, \dots\}$ with

$$Y_i = \begin{cases} X_i \setminus \{m_i\} & \text{if } m_i \in X_i, \\ X_i \cup \{m_i\} & \text{else,} \end{cases}$$

and finally $Z = \{z_i := d(X_i, Y_i)/|M| \mid X_i \in \mathscr{X}, Y_i \in \mathscr{Y}\}$ as the set of pairwise closure Hamming distances. The network is trained by showing it the binary encodings of X_i, Y_i, and z_i. Note, that the values of Z are normalized. We will evaluate this setup in Sect. 3.5.2 on different data sets and for relevant notions of FCA.

3.4.2 Object2Vec and Attribute2Vec

The last section adumbrates that word2vec-based approaches may be unsuitable for embeddings of closure operators. However, we still adapt word2vec to the domain of formal concept analysis and study its aptitude in a different area. Namely, we introduce an approach to compute embeddings of objects and attributes that incorporates nearness. Here, nearness of two objects or attributes refers to the amount of concepts that include both.

The idea of adapting word2vec to non-text mining problems is a common approach these days. Particular examples for that are node2vec [13] and *deepwalk* [25]. In the realm of networks, it was shown that SG based architectures for node embeddings can beat former approaches that use classic graph measures. They significantly enhanced node classification and link prediction [13, 25] tasks. To do so, they interpret nodes as words for their vocabulary, use random walks through the graphs to generate "sentences", and then employ word2vec.

Analogously, we transfer word2vec to the realm of formal concept analysis. In the following we present an approach to use the concepts of a given formal context to generate embeddings of the object set or attribute set. Referring to its origin, we name our novel methods *object2vec* and *attribute2vec*, respectively. Since both methods will emerge to be dual to each other for obvious reasons we only consider object2vec in the following. The basic idea is to interpret two objects to be more close to each other, if they are included in more concept extents together. Hence, the set of extents of a formal context is used to generate a low dimensional embedding of the object set G.

In the following we explain how to adapt the CBOW and the SG architecture to the realm of formal concept analysis. We show how to generate (multi-) sets of training examples from a given formal context. As an analogon for target word context words we introduce target object and context object sets. From this we can draw pairs as already done in CBOW and SG.

3.4.2.1 SG and CBOW in the Realm of Object2Vec

Let $\mathbb{K} := (G, M, I)$ be a (finite) formal context. The vocabulary is given by $G = \{g_1, \ldots g_n\}$. Furthermore, let be $\phi : G \to \mathbb{R}^n, g_i \mapsto e^i$ the one-hot encoding of our vocabulary (objects). We derive our *training examples* from the set

$$T(\mathbb{K}) := \{(a, A \setminus \{a\}) \mid a \in A, |G| > |A| > 1, \exists B \subseteq M : (A, B) \in \mathfrak{B}(\mathbb{K})\},$$

where every element is a pair of a *target object* and some extent in which a is element of. More specifically, we remove a from this extent. We interpret then $A \setminus \{a\}$ as the *object context set*. The word "context" here refers to the word2vec approach and is not be confused with "formal context". Note that we do not generate any training examples from the concept (G, G') since the extent G does not provide any information about the formal context.

The Skip-gram Architecture for Object2Vec. In the SG model, the input and output training pairs generated from a target object and an object context set, i.e., the elements $(t, C) \in T(\mathbb{K})$, are given by:

$$T_{SG}(t, C) := \{(\phi(t), \phi(c)) \mid c \in C\}. \tag{3.2}$$

Using this it is possible for some pairs $(t_1, C_1), (t_2, C_2) \in T(\mathbb{K})$ where we have $(t_1, C_1) \neq (t_2, C_2)$ that $T_{SG}(t_1, C_1) \cap T_{SG}(t_2, C_2) \neq \emptyset$. Hence, samples can be generated multiple times in this setup. We account for this in our modeling, as the reader will see in the presentation of the algorithm. To give an impression of our modeling we furnish the following example.

Example 3.1 Consider a classical formal context from [11], called "Living beings and Water", which we depicted in Fig. 3.4. We map the objects with the one-hot encoding: $\phi : G \mapsto \mathbb{R}^8$, with $\phi(a) = e^1, \phi(b) = e^2, \ldots, \phi(h) = e^8$. Using this we easily find a training sample from T_{SG} which is generated two times. Consider the concepts $(\{a, f, g\}, \{4, 5, 6\})$ and $(\{f, g, h\}, \{5, 6, 7\})$. The first concept generates the pairs of target object and object context sets $(a, \{f, g\})$ as well as $(f, \{a, g\})$ and $(g, \{a, f\})$. The second formal concept generates the pairs $(f, \{g, h\}), (g, \{f, h\})$ and $(h, \{f, g\})$. If we train in the SG architecture we derive from the pair $(f, \{a, g\})$ the training examples (e^6, e^1) and (e^6, e^7). Also, from the pair $(f, \{g, h\})$ we derive the examples $(e^6, e^7), (e^7, e^8)$. Hence, the training example (e^6, e^7) is shown to be drawn at least twice per epoch.

The CBOW Architecture for Object2Vec. Analogously to the cases of CBOW in word2vec we will use for object2vec a notion of "middle point" for object context sets. More specifically, for a pair of target object and object context set $(t, C) \in T(\mathbb{K})$ the training example is derived as follows:

$$T_{CBOW}(t, C) := \left(\frac{1}{|C|} \sum_{c \in C} \phi(c), \phi(t) \right) \tag{3.3}$$

	1	2	3	4	5	6	7	8	9
a				×	×	×			×
b	×	×	×			×			
c	×	×		×		×		×	
d	×	×	×	×		×			
e	×		×			×			
f				×	×	×	×		
g			×	×	×	×	×		
h			×		×	×	×		

Fig. 3.4: Formal context of the classical "Living beings and Water" example from [11]

Hence, in the CBOW model the set of all training examples is given by

$$\mathrm{T_{CBOW}}(\mathbb{K}) := \{\mathrm{T_{CBOW}}(t,C) \mid (t,C) \in \mathrm{T}(\mathbb{K})\}.$$

Lemma 3.1 *The map* $\mathrm{T_{CBOW}} : \mathrm{T}(\mathbb{K}) \to \mathbb{R}^2, (t,C) \mapsto \left(\frac{1}{|C|} \sum_{c \in C} \phi(c), \phi(t)\right)$ *is injective.*

Proof Let E_n be the set of standard basis vectors. We first show that

$$f : 2^{E_n} \to \mathbb{R}^n, E \mapsto \frac{1}{|E|} \sum_{e \in E} e$$

is injective. Let $E_1, E_2 \in 2^{E_n}$ with $E_1 \neq E_2$ where we assume $E_1 \nsubseteq E_2$ w.l.o.g.. Hence, we find $e^i \in E_1 \setminus E_2$. It then follows $f(E_1)_i > 0 = f(E_2)_i$, therefore $f(E_1) \neq f(E_2)$ and f is injective. The function $\phi : G \to E_n$ is also injective, so the map $\Phi : 2^G \to 2^{E_n}, A \mapsto \Phi(A) := \phi[A] := \{\phi(a) \mid a \in A\}$ is also injective. For all $(t,C) \in \mathrm{T}(\mathbb{K})$ the equality

$$\mathrm{T_{CBOW}}(t,C) = (f(\Phi(C)), \phi(t))$$

holds. Hence, $\mathrm{T_{CBOW}}$ is injective.   □   □

It follows that the modeling of training samples as set (cf. Eq. (3.3)) is appropriate since no training example is derived multiple times from $\mathrm{T}(\mathbb{K})$.

Order of the training examples. Let $\mathbb{K} := (G,M,I)$ be a formal context. We want to embed the set of objects. Since we model our training examples as sets (with frequency in the case of SG) we need to discuss how to construct a traversable list of training examples for our training procedures. This is not necessary in word2vec where the order is given naturally by the order of the given text. We propose to generate the traversable list L in the following manner:

1. For all extents A of $\mathbb{K}$, construct a list L_A that consists of all elements of A. The order in the list L_A should be random.

Algorithm 1: The pseudocode of object2vec. The algorithm takes a formal context and an option determining Skip-gram or Continous Bag of Words. It returns a list of pairs to train the neural network

---

Input : a formal context (G, M, I) and type $\in \{SG, CBOW\}$
Output A list L of training examples.
:

1 $L \leftarrow [\,]$
2 $L_{ext} < -$ list of extents of (G, M, I) in randomized order (excluding G).
3 **forall** A *in* L_{ext} **do**
4      $L_A \leftarrow$ list(A), with randomized order.
5   **forall** o *in* L_A **do**
6   **if** *type* $=$ SG **then**
7   **forall** $ŏ$ *in* L_A: **do**
8   **if** $o \neq ŏ$ **then**
9   add $((\phi(o), \phi(ŏ)))$ to L

10   **if** *type* $=$ CBOW *and* $|A| > 1$ **then**
11   add $((\frac{1}{|L_A|-1} \sum_{ŏ \in l_a, o \neq ŏ} \phi(ŏ), \phi(o)))$ to L

12 **return** L

---

2. Construct a list L_{ext} that consists of all lists L_A. The order f L_{ext} should be randomized.
3. For each L_A such that $|A| > 1$ use Eq. (3.3) or Eq. (3.2) to add the training examples to the list L of all training examples.

We present an algorithmic representation of this course of action in Algorithm 1.

3.5 Experiments

This section contains experimental evaluations for both our research directions. We conduct our experiments on three different data sets. We depict the statistical properties of these data sets in Table 3.1. A detailed description of each follows.

wiki44k. The first data set we use in this work is the wiki44k data set taken from [16] and then adapted by [15]. It consists of relational data extracted from Wikidata in December 2014. Even though it is constructed to be a dense part of the Wikidata knowledge graph, it is relatively sparse for a formal context.

Mushroom. The Mushroom data set [9, 29] is a well investigated and broadly used data set in machine learning and knowledge representation. It consists of 8124 mushrooms. It has twenty two nominal features that are scaled into 119 different binary attributes to form a formal context. The Mushroom data set, compared to wiki44k, is more dense, and even though it has a smaller number of objects, contains 10 times the concepts of wiki44k.

ICFCA. The ICFCA context consists of all authors that have a publication at the *International Conference on Formal Concept Analysis*(ICFCA). The attributes are given by all publications of these authors. By all publications, we denote all their works present in DBLP, not restricted to ICFCA proceedings. To generate the ICFCA context, we use the DBLP dump from 2019-08-01 which can be found at https:// dblp.uni-tricr.de/xml/. We exclude authors in the DBLP data that have the type "disambiguation" or "group". To exclude publication originating from editing etc (which do not indicate any co-authorship) we discard all publications that are not of the type "article", "inproceedings", "book" or "incollection". We also discarded all publications that are marked with an additional "publtype" such as "withdrawn", "informal" and "informal withdrawn". Note that this also excludes works that are solely published on preprint servers. This modeling results in a formal context with 351 objects and 12614 attributes. However, as later indicated in the experiments, for the neural network training we will use only a part of that formal context. This part is derived by omitting all publications after 2015 and then considering the largest component. The specifications of the resulting formal context ICFCA* are depicted in Table 3.1. The ICFCA data set is available in the `conexp-clj`[2] software [14] for FCA hosted on GitHub. By the nature of being based on a publication network, it is very sparse and contains only 878 concepts.

3.5.1 Object2Vec and Attribute2Vec

We evaluate our new approaches object2vec and attribute2vec with two distinct experiments. First, we will study embeddings in the realm of link prediction. For this, we investigate a self-created publication network as described in the last section, called ICFCA. In this formal context, consisting of authors as objects and publications as attributes, the incidence relation is then given by *g is author of m*. Link prediction tasks can be split into two categories: decide in a network which links are missing or predict from a given temporal network snapshot which new links will occur in the future. In general this experiment evaluates the ability of object2vec to enhance link prediction.

In our second experiment we present a task that is of more general interest to formal concept analysis concerned research. We investigate a correspondence between the canonical base of implication $\mathscr{L}$ for a given formal context (G, M, I) and our embedding methods. In particular, we cluster the set of attributes M based on attribute2vec using a partitioning procedure and obtain a clustering $\mathscr{C}$. We then count the number of implications $A \to B$ from $\mathscr{L}$ that are in subset relation with a cluster, i.e., $A \cup B \subseteq C$ for some cluster $C \in \mathscr{C}$. With that we evaluate to which extent attribute2vec embeddings are able to reflect parts of the implicational structure from a formal context.

---

[2] See https://github.com/tomhanika/conexp-clj.

	Wiki44k	Mushroom	ICFCA*
Number of objects	45021	8124	263
Number of attributes	101	119	8442
Density	0.04	0.19	0.005
Number of concepts	21923	238710	680
Mean attributes per concept	7.01	16.69	33.28
Mean objects per concept	109.47	91.89	2.51
Size of the canonical base	7040	2323	?

Table 3.1: Comparison of the different data sets used in this work. For ICFCA we do only indicate the specification of the context as used for the training in the link prediction model. To compute the canonical base of the ICFCA data set is not feasible for the equipment at our research group

3.5.1.1 Link Prediction Using Object2Vec

Network embedding techniques like the prominent node2vec approach have proven their capability to predict links in huge networks [13]. Even though these methods employ low dimensional embeddings for their computations, the actual employed dimension is still incomprehensibly high for human understanding, i.e., more than 100. The realm of formal concept analysis is especially interested in interpretable and explainable methods. Hence, we focus on embeddings into $\mathbb{R}^2$ and $\mathbb{R}^3$.

Using the aforedescribed ICFCA data set our goal for the link prediction task is as follows. For learning an embedding we restrict ICFCA to the largest connected component we discover after we omitted all publications later than 2015. The prediction task then is to find future co-authorships. More specifically, we predict these co-authorships in the time interval from 2016-01-01 until 2019-08-01. We compare the introduced object2vec using both architectures, i.e., CBOW and SG, and compare the results with link prediction computed via node2vec. We may note that the node2vec embeddings are conducted in two and three dimensions as well. Our precise experimental pipeline looks as follows.

Compute an embedding via object2vec We first compute an embedding of the formal context with the object2vec approach. For the training of the neural network we use a learning rate with linear decay. We empirically set the starting learning rate to 1.0 and train for 200 epochs. We repeat the embedding process for 30 times.

Compute the node2vec embedding The node2vec embedding uses the parameters as used in [13], i.e., 10 walks per node with length 80 and a window size of 10. We use the standard learning rate of 0.025 as well as 1.0 for comparability reasons with respect to object2vec. However, we do not report the results for 1.0 since they were worse for the two and three dimensional case. We also repeat the embedding process for 30 times. As in [13], the procedure parameter p and q of node2vec are chosen by grid search in $\{0.25, 0.5, 1, 2, 4\}$.

Edge vector generation Since link prediction is concerned with edges we need to focus on the edge set. To generate the edge vectors from the node embedding we use the component-wise product of two node vectors. In [13] it was shown that this practice is favorable. Furthermore, in the special case of bibliometric link prediction this is the common approach, cf. [10].

Training examples Our learning procedure demands for training examples. The positive training examples are the edges in the co-author graph until (including) 2015. For each positive example, we select one negative example, i.e., two randomly picked nodes without an edge connecting them. This approach leads to 1278 training examples.

Test examples The positive examples are the author pairs which have an edge after 2016-01-01, but not before. For each such pair, we choose one negative example of two authors that neither co-authored before or after 2016. Using this we obtain 84 test examples, half of them positive.

Classification For the binary classification problem, we employ logistic regression. In particular, we use the implementation provided by *Scikit Learn* [23]. Logistic regression is often used in classification emerging from embeddings in the realm of word2vec, e.g., in [13, 25]. To determine the C parameter of the classifier, we do a grid search over $\{10^{-3}, 10^{-2}, \ldots, 10^2\}$.

Embedding	Dim	Recall		Precision		F1-score	
		Mean	Stdev	Mean	stdev	Mean	stdev
node2vec	2	0.56	0.14	0.60	0.07	0.57	0.09
O2V-SG	2	0.66	0.08	**0.65**	0.03	0.65	0.05
O2V-CBOW	2	**0.68**	0.09	0.64	0.04	**0.66**	0.06
node2vec	3	0.60	0.15	0.56	0.08	0.58	0.10
O2V-SG	3	0.70	0.08	0.62	0.04	0.66	0.06
O2V-CBOW	3	**0.73**	0.07	**0.65**	0.06	**0.69**	0.06

Table 3.2: The result of our classification experiment. We compare node2vec to object2vec with the Skip-gram architecture (O2V-SG) and with the object2vec Continuous Bag of Words architecture (O2V-CBOW). We display the mean value over the 30 rounds of the experiments and also present the sample standard deviation

The results of our experiments are depicted in Table 3.2. We observe that in all three indicators, i.e., recall, precision and F1-Score, node2vec is dominated by the object2vec approach. Furthermore, we see that the CBOW architecture performs better compared to SG in almost all cases. However, this benefit is small and well covered in the standard deviation. Finally, we find that embeddings in three dimensions perform generally better than in two.

3.5.1.2 Clustering Attributes with Attribute2Vec

In FCA, implications on the set of attributes of a formal context are of major interest. While computing the canonical base, i.e., the minimal base of the implicational theory of a formal context, is often infeasible, one could be interested in implications of smaller attribute subsets. This leads naturally to the question of how to identify attribute subsets that cover a large part of the canonical base, as explained in the beginning of Sect. 3.5.1. In detail, the resulting task is as follows: Let (G, M, I) be a formal context and $\mathscr{L}$ the canonical base (G, M, I). Using a simple clustering procedure (in our case k-means), find for a given $k \in \mathbb{N}_0$ a partitioning of M in k clusters such that the ratio of implications completely contained in one cluster (cf. Sect. 3.5.1) is as high as possible. We additionally constrain this task by limiting clusterings in which the largest cluster is significantly smaller than $|M|$.

We investigate to which extent our proposed approach attribute2vec maps attributes closely that are meaningful for the aforementioned task. We conduct this research by computing an embedding via attribute2vec and run the k-means clustering algorithm on top of it. We evaluate our approach on the introduced wiki44k data set. Again, we refer the reader to the collected statistics of this data set in Table 3.1. The experimental pipeline looks as follows.

Applying attribute2vec on wiki44k We start by computing two and three dimensional vector embeddings of the wiki44k attributes using attribute2vec. We employ both architectures, SG and CBOW. Here we use again the starting learning rate of 1.0 with linear decay. In contrast to the embeddings of the ICFCA data set we find that 5 training epochs are sufficient for stabilization of the embeddings.

K-means clustering We use the computed embedding to cluster our attributes with the k-means algorithm. As implementation we rely on the Scikit Learn software package. For the initial clustering, we use the so called "k-means++" technique by [3]. The method from Scikit Learn runs internally for ten times with different seeds and returns the best result encountered with respect to the costs of the clusters. We choose k from $\{2, 5, 10\}$. We denote the resulting clustering with $\mathscr{C}$.

Computation of the intra-cluster implications An implication drawn from the canonical base, i.e., $A \to B \in \mathscr{L}$, is called *intra-cluster* if there is some $C \in \mathscr{C}$ such that $A \cup B \subseteq C$. The canonical base of wiki44k has the size 7040. For a clustering $\mathscr{C}$ we compute the ratio of intra-cluster implications.

Repetition We repeat the steps above for 20 times. Hence, we report the mean as well as the standard deviation of all the results.

Baseline clusterings To evaluate the ratios computed in the last step we use the following baseline approaches. As a first baseline we make use of a random procedure. This results in a random clustering of the attribute set. Using an arbitrary random clustering with respect to cluster sizes is unreasonable for comparison. Hence, for each k-means clustering obtained above we generate 50 random clusterings of the same size and the same cluster size distribution. For those we also compute the intra-cluster implication ratio.

# clusters		2		5		10	
Type	dim	Mean	stdev	Mean	stdev	Mean	stdev
Skip-gram	2	0.1608	0.0031	0.0703	0.0122	0.0069	0.0004
Random	2	0.0534	0.0412	0.0084	0.0088	0.0010	0.0007
Skip-gram	3	**0.3217**	0.0005	**0.1028**	0.0218	**0.0080**	0.0001
Random	3	0.0219	0.0107	0.0036	0.0027	0.0007	0.0004
Naive clustering	-	0.0158	0.0000	0.0055	0.0042	0.0035	0.0002

Table 3.3: Results of the clustering task. For $k \in \{2,5,10\}$ we display the mean and standard deviation of the ratio of intra-cluster implications. For the dimensions 2 and 3 we show the mean and sample standard deviation values for the Skip-gram architecture and the random clusters with same cluster size distributions as the corresponding Skip-gram clusterings. We also compared our method with the "naive" clustering approach

Naive k-means clustering As second baseline we envision a more sophisticated procedure. We call this the "naive" clustering approach. In this setting we encode an attribute m through a binary vector representation using the objects from $\{m\}'$ as described in Sect. 3.4.1. We then run twenty rounds of k-means and compare the results with the attribute2vec approach. For comparison we use again $k \in \{2,5,10\}$.

We display our observations from this experiment in Table 3.3. In there we omit results for the CBOW architecture since they do not exceed the results obtained in the random baseline approach. As our main result we find that the Skip-gram architecture in three dimensions achieves the best intra-cluster ratio. More specifically, SG outperforms for all cluster sizes and all dimension the baseline approach and the naive clustering approach by a large margin. This margin, however, is smaller in the two dimensional case compared to the three dimensional try. For smallest investigated clustering size, i.e., two, naive clustering performs worse than the random baseline. We can report for our experiments the following average maximum cluster sizes. For dimension three we have 54.5 attributes for $k = 2$, 33.2 attributes for $k = 5$, and 14.9 attributes for $k = 10$. Finally, we also spot for clustering sizes five and ten that the naive clustering does operate better than the random baseline.

3.5.1.3 Discussion

In both our experiments we find that all embedding procedures perform better in dimension three than in dimension two. This is not surprising since a higher embedding dimension possess a higher degree of freedom to represent structure. Furthermore, in both experiments we can show that the object2vec and attribute2vec approaches do succeed and outperform the competition.

The first experiment reveals some particularities. We find that our embedding approach has a big advantage over the also considered node2vec procedure. For the later method one has to perform additional parameter tuning for the p and q parameters [13] to obtain the presented results. Not doing so leads to worse performance. The object2vec embedding procedure, as defined in Sect. 3.4.2, needs no parameters for the training example generation. In addition to that, the set of computed training examples is deterministic. However, these positive properties come with a high computational cost, i.e., the necessity of computing the set of formal concepts. For the sake of completeness we also compute high dimensional embeddings. When applying node2vec with embedding dimension one hundred we find the results outperforming the so far reported. Still, since such embeddings conflict with the goal in this work, i.e., the human interpretability and explainability of embeddings, we discard them.

The second experiment also unraveled different properties of our novel embedding technique. We witness that the number of learning epochs is much smaller compared to the first experiment. We suspect that this can be attributed to the higher average number of attributes per intent in wiki44k compared to the average number of objects per extent in ICFCA. Furthermore, we think there is an influence by the fact that the absolute number of intents in wiki44k is greater than the absolute number of extents in ICFCA. As application for our attribute2vec approach we envision the computation of parts of the canonical base of a formal context. Taking the average maximum cluster sizes into account we claim that this application is reasonable. We do not consider the then necessary computation of all formal concepts as a disadvantage. The computation of the canonical base is in general far more complex than computing the set of all concepts and our embedding. We are surprised that the naive clustering baseline performs not significantly better than the random baseline. We presume that k-means clustering applied to the binary representation vectors is not useful to reflect implicational knowledge in the embedding space. Hence, we admit that more powerful baseline comparisons may be considered here. However, so far we are not aware of less computationally demanding ones with respect to object2vec and attribute2vec respectively.

3.5.2 FCA Features Through Closure2Vec

To evaluate the embeddings produced by closure2vec we introduce two FCA related problems: computing the covering relation of a concept lattice and computing the canonical base for a given formal context. The intention here is to rediscover structural features from FCA in low dimensional embeddings. We choose for the dimension two and three in order to respect our overall goal for human interpretability and explainability. We test two different functions δ for the distance between the output layers O_L, O_L', cf. Sect. 3.4.1.5. More specifically, we employ the Euclidean distance and the cosine distance. We conduct our experiments on two larger than average sized formal contexts. Precisely, we test the Wiki44k [15, 16] and the well investigated

Mushroom data set [9, 29]. A comparison of the statistical properties of data sets we use is depicted in Table 3.1.

3.5.2.1 Distance of Covering Relation

For this experiment we compute first the set of all concepts $\mathfrak{B}$ of a given formal context $\mathbb{K}$. Using the concept order relation $<$ as introduced in Sect. 3.3.1 a *covering relation* on $\mathfrak{B}$ is given by: $\prec \subseteq (\mathfrak{B} \times \mathfrak{B})$ with $A \prec B$, if and only if $A < B$ and there is no $C \in \mathfrak{B}$ such that $A < C < B$. The covering relation is an important tool in *ordinal data science*. Elements of the covering relation are essential for investigating and understanding order relations and order diagrams. However, in the case of large formal contexts computing the covering relation of the concept lattice can get computationally expensive, as this problem is linked to the transitive reduction of a graph [2].

The experimental setup now is as follows. First we train the neural network architecture as introduced in Sect. 3.4.1.5. Hence, an input element is a binary encoded attribute X_i, another binary encoded attribute set Y_i of size $|X_i| \pm 1$, and the closure Hamming distance of them. In our experiment we fix $|X_i|$ to be four or less. Furthermore, we train the network over five epochs using the learning rate 0.001, with batch size 32, and mean-squared-error as loss function.

To evaluate the structural quality of the obtained embedding we computed the covering relation of the concept lattices using 1000 threads on highly parallelized many-core systems, which took about one day. In the following, we compare the distances between pairs of concepts in covering relation against concepts that are not in covering relation. The results of these experiments are depicted in Table 3.4. For all embeddings the expected distance for two concepts in covering relation is significantly smaller than the expected difference of two concepts not in covering relation. This is true for both data sets. However, the observed effect is more notable for the wiki44k data set. The Euclidean distance outperforms the cosine distance in all experiments using two and three dimensions.

3.5.2.2 Distance of Canonical Bases

In this experiment we look at the canonical base of implications for a formal context and try to rediscover this canonical base in the computed embedding. The experiment consists of two different parts. The first part has the following setup. Take an implication of the canonical base, i.e., take (P, C) where $P \subseteq M$ is the premise and $C \subseteq M$ is the conclusion. For such an implication, construct all single conclusion implications (P, c) where $c \in C$. Then, compute the distance (with the same distance functions as used for the embedding process) of P and the embedding for all c. Additionally, also embed all $m \in M \backslash C$. Essentially, by doing so, we embedded all $m \in M$ using our embedding function. We do this for all elements of the canonical base.

Wiki44k:

Dim 2		Mean: Std.:	Dim 3		Mean: Std.:
Euc:	CR:	0.17 0.14	Euk:	CR:	0.16 0.15
	Non-CR:	0.71 0.59		Non-CR:	1.54 1.41
Cos:	CR:	0.63 0.33	Cos:	CR:	0.15 0.27
	Non-CR:	0.99 0.71		Non-CR:	0.36 0.43

Mushrooms:

Dim 2		Mean: Std.:	Dim 3		Mean: Std.:
Euc:	CR:	0.14 0.41	Euc:	CR:	0.13 0.29
	Non-CR:	0.51 0.38		Non-CR:	0.49 0.41
Cos:	CR:	0.49 0.43	Cos:	CR:	0.05 0.12
	Non-CR:	0.96 0.74		Non-CR:	0.18 0.22

Table 3.4: Distance between concept pairs, that are in covering relation (CR) and that are not in the covering relation (Non-CR)

Wiki44k:

Dim 2		Mean: Std.:	Dim 3		Mean: Std.:
Euc:	S-Imp:	0.94 0.28	Euc:	S-Imp:	0.93 0.29
	Non-S-Imp:	0.73 0.40		Non-S-Imp:	0.83 0.47
	Imp:	0.44 0.33		Imp:	0.60 0.45
	Non-Imp:	0.51 0.48		Non-Imp:	0.65 0.55
Cos:	S-Imp:	1.00 0.69	Cos:	S-Imp:	0.69 0.53
	Non-S-Imp:	1.00 0.67		Non-S-Imp:	0.90 0.43
	Imp:	1.01 0.70		Imp:	0.93 0.56
	Non-Imp:	1.01 0.70		Non-Imp:	0.96 0.58

Mushrooms:

Dim 2		Mean: Std.:	Dim 3		Mean: Std.:
Euc:	S-Imp:	0.95 0.41	Euc:	S-Imp:	0.86 0.34
	Non-S-Imp:	0.45 0.32		Non-S-Imp:	0.42 0.27
	Imp:	0.70 1.01		Imp:	0.57 0.66
	Non-Imp:	1.02 0.98		Non-Imp:	0.88 0.68
Cos:	S-Imp:	1.00 0.65	Cos:	S-Imp:	1.05 0.36
	Non-S-Imp:	1.00 0.65		Non-S-Imp:	1.02 0.35
	Imp:	1.00 0.69		Imp:	0.88 0.53
	Non-Imp:	0.99 0.69		Non-Imp:	0.89 0.55

Table 3.5: Distances of implication premises and conclusions for singleton implications (S-Impl) and implications (Impl) in the computed embeddings

Now compute for all implications from the canonical base the following distances. First, the distances between a premise P and all its singleton conclusions c of (P, C). Secondly, the distances between P and $m \in M \setminus C$. Equipped with all these distances we try to detect a structural difference in favor of the embedded implications in contrast to other combinations of attribute sets, i.e., pairs of premises P and $m \in M \setminus C$. When using the cosine distance function we observe minimal structural difference. However, when using the Euclidean distance function we detect a significant structural difference. In particular, for pairs of (P, c) with $c \in C$ the mean of the distances is significantly higher than the mean distance of the distances for some premise P with all singleton sets $m \in M \setminus C$. The observation is even stronger in the case of the Mushroom data set when compared to wiki44k. The results are depicted in Table 3.5 by the rows S-Imp (for the combinations (P, c)) and Non-S-Imp (for combinations (P, m)).

For the second part we embed both attribute sets, i.e., the premise P and the conclusion C, for an implication from the canonical base. For every pair we compute the distance of P and C and compare them to the distance between two randomly generated attribute sets X, Y with $|X| = |P|$ and $|Y| = |C|$. The later is set for reasons of comparability. As shown in Table 3.5 we detect again structural differences for the considered implications and the randomly generated sets using average distances as features. In fact, the distance between the two randomly generated sets is on average larger than the average distance between premises and conclusions from implications drawn from the canonical base.

3.5.2.3 Discussion

In both experiments concerning the closure system embedding we are able to rediscover and infer conceptual structures in the embeddings. In general, we find that it is favorable to use for the Euclidean distance case the squares of the closure Hamming distances as output, i.e., for z_i. Overall we discovered a significant bias in distances of embedded concepts that are in covering relation. This signal is even stronger for the wiki44k data sets. We suspect that this can be attributed to the lesser density of this data set compared to Mushroom. The observations can be exploited naturally for mining covering relations, or important parts of those, from embedded concept lattices.

For the second experiment we can report that the neural network embedding of parts of the closure system allows for rediscovering implicational structures. Since we trained our neural network on attribute sets of size four and smaller, we were interested in the number of closures our algorithm encounters. For both data sets we can report that this number is approximately 10% of all closures.

The described behavior differs between the experiment where we train the network using whole implications, i.e., premise and conclusions, and single-conclusion implications. In cases where an attribute is element of the conclusion of some canonical base implication the distance to the premise set is significant. At this point we are unable to provide a rational for that. The same goes for the second part of the experiment where we compare premise-conclusion pairs of canonical base implications with randomly generated pairs of attributes having the same sizes. At this point we are not aware how this observation can improve the computations of the canonical base. This would need a more fundamental investigation of bases of impli-

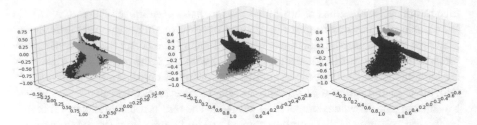

Fig. 3.5: Embedding of all concepts of the mushroom dataset into three dimensions. The coloring is done as follows. Left: The edible mushrooms are green, the non-edible mushrooms are red. Middle: The mushrooms with a broad gill are green, the mushrooms with a narrow gill are red. Right: The mushrooms with a crowded gill spacing are green and the mushrooms with a distant gill spacing are red

cational theories with respect to closure system embeddings. For example, one has to investigate if the observed effect is also true for other kinds of bases, e.g., direct basis [1]. As a final remark we report that the Euclidean distance performed in all our experiments better than the cosine distance for both problem settings.

3.5.2.4 Empirical Structural Observations

Additionally to the two experiments above we want to provide some insights we discovered for conceptual structures in our embeddings. We note that concepts sharing attributes seem to result in meaningful clusters. To see this one can consider Fig. 3.5. In there we see the same embedding of all formal concepts of the Mushroom data set in three dimensions for three times. In each case we colored different sets of concepts with red and green. In the first (Fig. 3.5, left) we depict with red the not edible mushrooms and with green the edible ones. Even though we employed a very low dimensional embedding, we can still visually identify the two different classes. Hence, our embedding approach preserved some structure. The same seems true for the other depictions in which we colored broad gill versus narrow gill and crowded gill spacing versus distant gill spacing. Therefore, we are confident that our approach for low dimensional embeddings of closure systems using neural networks is beneficial. Moreover, as this empirical study shows, the low dimensional representation is still visitable by a human data analyst.

3.6 Conclusion

In this work we presented fca2vec, a first approach for modeling data and conceptual structures from FCA to the realm of embedding techniques based on word2vec. Taken together, the ideas in this paper outline an appealing connection between formal concepts, closure systems, low dimensional embeddings, and neural networks learning procedures. We are confident that future research may draw on the demonstrated first steps, in particular object2vec, attribute2vec and learning closure operator representations. In our investigation we have found convincing theoretical as well as experimental evidence that FCA-based methods can profit from word2vec-like embedding procedures. We demonstrated that closure operator embeddings that result from simple neural network learning algorithms may capture significant portions of the conceptual structure. Furthermore, we were able to demonstrate that the covering relation of the set of formal concepts may be partially extracted from a low dimensional embedding. Especially when employing conceptual structures in large and complex data this notion is an important step forward. Moreover, we were able to enhance the common embedding approach node2vec in low dimensional cases, i.e., dimension two or three.

All these results were achieved while obeying the constraint for human interpretable and/or explainable embeddings. Applying neural network learning procedures on large and complex data does not necessarily constitute a contradiction to explainability when combined with conceptual notions from FCA. However, our work clearly has some limitations. The ideas for object2vec and attribute2vec do require the computation of the concept lattice. In future work we will investigate if this obligation can be weakened through statistical methods. Despite this we believe that our work could be the standard framework for word2vec-like FCA approaches. As a next concrete application we are currently in the process of investigating genealogy graphs in combination with co-authorship networks. These multi-relational data sets are large and complex and do require novel methods, like fca2vec, to draw knowledge from them. Questions for the relation of particular nodes in such data sets may be answered through conceptual embeddings. In this context we do also take Resource Description Framework (RDF) structures into account. Ideas for embedding those is a state-of-the-art approach to knowledge graph structures. Hence, enhancing RDF embeddings using fca2vec as well as discovering conceptual structures in RDF is a fruitful endeavor. Finally, on a more technical note, we are interested in characterizing sets of formal context data. This would allow for particular representations of the closure operator, e.g., closure operators representable by affine maps.

Acknowledgements This work is partially funded by the German Federal Ministry of Education and Research (BMBF) in its program "Quantitative Wissenschaftsforschung" as part of the REGIO project under grant 01PU17012, and in its program "Forschung zu den Karrierebedingungen und Karriereentwicklungen des Wissenschaftlichen Nachwuchses (FoWiN)" under grant 16FWN016.

References

1. Adaricheva, K.V., Nation, J.B., Rand, R.: Ordered direct implicational basis of a finite closure system. Discrete Applied Mathematics **161**(6), 707–723 (2013)

2. Aho, A.V., Garey, M.R., Ullman, J.D.: The transitive reduction of a directed graph. SIAM Journal on Computing **1**(2), 131–137 (1972)
3. Arthur, D., Vassilvitskii, S.: k-means++: the advantages of careful seeding. In: N. Bansal, K. Pruhs, C. Stein (eds.) Proceedings of the Eighteenth Annual ACM-SIAM Symposium on Discrete Algorithms, SODA 2007, New Orleans, Louisiana, USA, January 7–9, 2007, pp. 1027–1035. SIAM (2007)
4. Belohlávek, R., Trnecka, M.: From-below approximations in boolean matrix factorization: Geometry and new algorithm. J. Comput. Syst. Sci. **81**(8), 1678–1697 (2015)
5. Belohlávek, R., Vychodil, V.: Discovery of optimal factors in binary data via a novel method of matrix decomposition. J. Comput. Syst. Sci. **76**(1), 3–20 (2010)
6. Bishop, C.M.: Pattern recognition and machine learning. Springer Science+Business Media (2006)
7. Caro-Contreras, D.E., Mendez-Vazquez, A.: Computing the concept lattice using dendritical neural networks. In: M. Ojeda-Aciego, J. Outrata (eds.) Proceedings of the Tenth International Conference on Concept Lattices and Their Applications, La Rochelle, France, October 15–18, 2013, *CEUR Workshop Proceedings*, vol. 1062, pp. 141–152. CEUR-WS.org (2013). URL http://ceur-ws.org/Vol-1062/paper12.pdf
8. Codocedo, V., Taramasco, C., Astudillo, H.: Cheating to achieve formal concept analysis over a large formal context. In: A. Napoli, V. Vychodil (eds.) Proceedings of The Eighth International Conference on Concept Lattices and Their Applications, Nancy, France, October 17–20, 2011, *CEUR Workshop Proceedings*, vol. 959, pp. 349–362. CEUR-WS.org (2011)
9. Dua, D., Graff, C.: UCI machine learning repository (2017). URL http://archive.ics.uci.edu/ml
10. Ganguly, S., Pudi, V.: Paper2vec: Combining graph and text information for scientific paper representation. In: J.M. Jose, C. Hauff, I.S. Altıngovde, D. Song, D. Albakour, S. Watt, J. Tait (eds.) Advances in Information Retrieval, pp. 383–395. Springer International Publishing, Cham (2017)
11. Ganter, B., Wille, R.: Formal Concept Analysis: Mathematical Foundations. Springer-Verlag, Berlin (1999)
12. Goldberg, Y., Levy, O.: word2vec explained: deriving Mikolov et al.'s negative-sampling word-embedding method. CoRR **abs/1402.3722** (2014)
13. Grover, A., Leskovec, J.: node2vec: Scalable feature learning for networks. In: B. Krishnapuram, M. Shah, A.J. Smola, C.C. Aggarwal, D. Shen, R. Rastogi (eds.) Proceedings of the 22nd ACM SIGKDD International Conference on Knowledge Discovery and Data Mining, San Francisco, CA, USA, August 13–17, 2016, pp. 855–864. ACM (2016)
14. Hanika, T., Hirth, J.: Conexp-clj - A research tool for FCA. In: D. Cristea, F.L. Ber, R. Missaoui, L. Kwuida, B. Sertkaya (eds.) Supplementary Proceedings of ICFCA 2019 Conference and Workshops, Frankfurt, Germany, June 25–28, 2019, *CEUR Workshop Proceedings*, vol. 2378, pp. 70–75. CEUR-WS.org (2019)

15. Hanika, T., Marx, M., Stumme, G.: Discovering implicational knowledge in wikidata. In: D. Cristea, F.L. Ber, B. Sertkaya (eds.) Formal Concept Analysis - 15th International Conference, ICFCA 2019, Frankfurt, Germany, June 25–28, 2019, Proceedings, *LNCS*, vol. 11511, pp. 315–323. Springer (2019)

16. Ho, V.T., Stepanova, D., Gad-Elrab, M.H., Kharlamov, E., Weikum, G.: Rule learning from knowledge graphs guided by embedding models. In: D. Vrandecic, K. Bontcheva, M.C. Suárez-Figueroa, V. Presutti, I. Celino, M. Sabou, L. Kaffee, E. Simperl (eds.) The Semantic Web - ISWC 2018 - 17th International Semantic Web Conference, Monterey, CA, USA, October 8–12, 2018, Proceedings, Part I, *LNCS*, vol. 11136, pp. 72–90. Springer (2018)

17. Kuznetsov, S.O., Makhazhanov, N., Ushakov, M.: On neural network architecture based on concept lattices. In: M. Kryszkiewicz, A. Appice, D. Slezak, H. Rybinski, A. Skowron, Z.W. Ras (eds.) Foundations of Intelligent Systems - 23rd International Symposium, ISMIS 2017, Warsaw, Poland, June 26–29, 2017, Proceedings, *Lecture Notes in Computer Science*, vol. 10352, pp. 653–663. Springer (2017)

18. LeCun, Y., Bengio, Y., Hinton, G.: Deep learning. Nature **521**, 436 (2015)

19. Mikolov, T., Chen, K., Corrado, G., Dean, J.: Efficient estimation of word representations in vector space. In: Y. Bengio, Y. LeCun (eds.) ICLR (Workshop Poster) (2013)

20. Mikolov, T., Sutskever, I., Chen, K., Corrado, G.S., Dean, J.: Distributed representations of words and phrases and their compositionality. In: C.J.C. Burges, L. Bottou, Z. Ghahramani, K.Q. Weinberger (eds.) Advances in Neural Information Processing Systems 26: 27th Annual Conference on Neural Information Processing Systems 2013. Proceedings of a meeting held December 5–8, 2013, Lake Tahoe, Nevada, United States, pp. 3111–3119 (2013)

21. Mnih, A., Hinton, G.E.: A scalable hierarchical distributed language model. In: D. Koller, D. Schuurmans, Y. Bengio, L. Bottou (eds.) Advances in Neural Information Processing Systems 21, Proceedings of the Twenty-Second Annual Conference on Neural Information Processing Systems, Vancouver, British Columbia, Canada, December 8–11, 2008, pp. 1081–1088. Curran Associates, Inc. (2008)

22. Nielsen, F.Å.: Wembedder: Wikidata entity embedding web service. CoRR **abs/1710.04099** (2017)

23. Pedregosa, F., Varoquaux, G., Gramfort, A., Michel, V., Thirion, B., Grisel, O., Blondel, M., Prettenhofer, P., Weiss, R., Dubourg, V., Vanderplas, J., Passos, A., Cournapeau, D., Brucher, M., Perrot, M., Duchesnay, E.: Scikit-learn: Machine learning in Python. Journal of Machine Learning Research **12**, 2825–2830 (2011)

24. Peng, H., Li, J., Song, Y., Liu, Y.: Incrementally learning the hierarchical softmax function for neural language models. In: S.P. Singh, S. Markovitch (eds.) Proceedings of the Thirty-First AAAI Conference on Artificial Intelligence, February 4–9, 2017, San Francisco, California, USA, pp. 3267–3273. AAAI Press (2017)

25. Perozzi, B., Al-Rfou, R., Skiena, S.: Deepwalk: online learning of social representations. In: S.A. Macskassy, C. Perlich, J. Leskovec, W. Wang, R. Ghani (eds.) The 20th ACM SIGKDD International Conference on Knowledge Discovery and Data Mining, KDD '14, New York, NY, USA - August 24 - 27, 2014, pp. 701–710. ACM (2014)
26. Ristoski, P., Rosati, J., Noia, T.D., Leone, R.D., Paulheim, H.: Rdf2vec: RDF graph embeddings and their applications. Semantic Web **10**(4), 721–752 (2019)
27. Rong, X.: word2vec parameter learning explained. CoRR **abs/1411.2738** (2014)
28. Rudolph, S.: Using FCA for encoding closure operators into neural networks. In: U. Priss, S. Polovina, R. Hill (eds.) Conceptual Structures: Knowledge Architectures for Smart Applications, 15th International Conference on Conceptual Structures, ICCS 2007, Sheffield, UK, July 22–27, 2007, Proceedings, *LNCS*, vol. 4604, pp. 321–332. Springer (2007)
29. Schlimmer, J.: Mushroom records drawn from the audubon society field guide to north american mushrooms. GH Lincoff (Pres), New York (1981)
30. Scott, D.: Measurement structures and linear inequalities. Journal of Mathematical Psychology **1**(2), 233 – 247 (1964)
31. Vrandecic, D., Krötzsch, M.: Wikidata: a free collaborative knowledgebase. Commun. ACM **57**(10), 78–85 (2014)
32. Wang, Z., Zhang, J., Feng, J., Chen, Z.: Knowledge graph embedding by translating on hyperplanes. In: C.E. Brodley, P. Stone (eds.) Proceedings of the Twenty-Eighth AAAI Conference on Artificial Intelligence, July 27 -31, 2014, Québec City, Québec, Canada, pp. 1112–1119. AAAI Press (2014)
33. Wille, U.: Representation of finite ordinal data in real vector spaces. In: H.H. Bock, W. Polasek (eds.) Data Analysis and Information Systems, pp. 228–240. Springer Berlin Heidelberg, Berlin, Heidelberg (1996)
34. Wille, U.: The role of synthetic geometry in representational measurement theory. journal of mathematical psychology **41**(1), 71–78 (1997)

Chapter 4
Analysis of Complex and Heterogeneous Data Using FCA and Monadic Predicates

Karell Bertet, Christophe Demko, Salah Boukhetta, Jérémy Richard, and Cyril Faucher

4.1 Introduction

Formal Concept Analysis (FCA) has been introduced in 1982 [17] then in the Ganter and Wille's work in 1999 [11]. FCA is issued from a branch of applied lattice theory that was introduced by Birkhoff in 1940 [4], then in the Barbut and Monjardet's work in 1970 [2].

Starting from a binary relation between a set of objects and a set of attributes, formal concepts are built as maximal sets of objects in relation with maximal sets of attributes, by means of derivation operators forming a Galois connection whose composition is a closure operator [3]. Concepts form a partially ordered set that represents the initial data called the concept lattice. This lattice has proved to be useful in many fields, e.g. artificial intelligence, knowledge management, data-mining, machine learning, etc.

Logical Concept Analysis [9] is a generalization of FCA in which sets of attributes are replaced by logical expressions. The power set of attributes mentioned by the Galois connection is replaced by an arbitrary set of formulas to which are associated a deduction relation (i.e. subsumption), and conjonctive and disjunctive operations, and therefore forms a lattice. LCA is the fundamental base of Abstract Conceptual Navigation [8] whose principle is to navigate in a conceptual space where places are logical concepts connected by navigation links.

K. Bertet (✉) • C. Demko • S. Boukhetta • J. Richard • C. Faucher
Laboratory L3i, La Rochelle University, La Rochelle, France
e-mail: karell.bertet@univ-lr.fr; christophe.demko@univ-lr.fr; salah.boukhetta@univ-lr.fr; jeremy.richard2@univ-lr.fr; cyril.faucher@univ-lr.fr

© The Author(s), under exclusive license to Springer Nature Switzerland AG 2022
R. Missaoui et al. (eds.), *Complex Data Analytics with Formal Concept Analysis*,
https://doi.org/10.1007/978-3-030-93278-7_4

For some complex data such as sequences or graphs, there are dedicated algorithms and tools to mine these data. Sequence mining is a topic of data mining which aims at finding patterns in a dataset of sequences that appears more frequently, with maximal common subsequences as patterns in CloSpan [19], prefix as patterns in PrefixSpan [14], etc. ... Graph mining aims at finding frequent common subgraphs in a dataset of graphs, with for example Gspan [18] and Plagram [15] for planar graphs.

Whether sequence mining, graph mining, pattern structures, logical concept analysis, ... these dedicated or generalized approaches allow to mine complex data, but often with a large number of generated patterns and unreasonably large space description. They are often untractable [12], uneasy to interpret, that hinders their ability to provide readability and explanation of the data. Moreover, these approaches do not allow an easy management of heterogeneous datasets where several kinds of characteristics describe data.

In a recent article[7], we have published a new algorithm, NEXTPRIORITYCONCEPT, inspired from pattern structures that locally computes the description of objects, in order to obtain well-suited descriptions of the objects, and smaller lattices. Our algorithm extends the computation of concepts to heterogeneous and complex datasets, where attributes are monadic predicates for each kind of data. Such a generic use of predicates regardless of the data allows a generic implementation of our algorithm which, mixed with a system of plugins, makes it possible to integrate new kinds of data. GALACTIC (GAlois LAttices, Concept Theory, Implicational systems and Closures) is a development platform of our algorithm allowing easy integration of new plugins.

In a first section, we present the NEXTPRIORITYCONCEPT algorithm and the GALACTIC development platform. In a second section, we show how this tool can be used for boolean, categorized, numerical, character string and sequential data on well-known examples of literature.

4.2 The NEXTPRIORITYCONCEPT Algorithm

4.2.1 Formal Concept Analysis

4.2.1.1 Concept Lattice

Let $\langle G, M, I \rangle$ a *formal context* where G is a non-empty set of objects, M is a non-empty set of attributes and $I \subseteq G \times M$ is a binary relation between the set of objects and the set of attributes. Let $(2^G, \subseteq) \xleftrightarrow[\alpha]{\beta} (2^M, \subseteq)$ be the corresponding *Galois connection* where:

- $\alpha : 2^G \to 2^M$ is an application which associates a subset $B \subseteq M$ to every subset $A \subseteq G$ such that $\alpha(A) = \{b : b \in M \wedge \forall a \in A, aIb\}$;

- $\beta : 2^M \to 2^G$ is an application which associates a subset $A \subseteq G$ to every subset $B \subseteq M$ such that $\beta(B) = \{a : a \in G \land \forall b \in B, aIb\}$.

A concept is a pair (A,B) such that $A \subseteq G$, $B \subseteq M$, $B = \alpha(A)$ and $A = \beta(B)$. The set A is called the *extent*, whereas B is called the *intent* of the concept (A,B). There is a natural hierarchical ordering relation between the concepts of a given context that is called the subconcept-superconcept relation:

$$(A_1,B_1) \leq (A_2,B_2) \iff A_1 \subseteq A_2 (\iff B_2 \subseteq B_1)$$

The ordered set of all concepts makes a complete lattice called the *concept lattice* of the context, that is, every subset of concepts has an infimum (meet) and a supremum (join).

4.2.1.2 Pattern Structures

Pattern structures [10] extends FCA to deal with non binary data. Formally, a pattern structure [10] is a triple $(G, (\mathcal{D}, \sqcap), \delta)$ where G is a set of objects, $(\mathcal{D}, \sqcap)$ is a meet-semi-lattice of potential objects descriptions, and $\delta : G \to \mathcal{D}$ associates each object with its description. Elements of $\mathcal{D}$ are ordered with the subsumption relation $\sqsubseteq$. Let $(2^G, \subseteq) \xleftrightarrow[\alpha_{\mathcal{D}}]{\beta_{\mathcal{D}}} (\mathcal{D}, \sqsubseteq)$ be the corresponding Galois connection where:

- $\alpha_{\mathcal{D}} : 2^G \to \mathcal{D}$ is defined for $A \subseteq G$ by $\alpha_{\mathcal{D}}(A) = \sqcap_{g \in A} \delta(g)$.
- $\beta_{\mathcal{D}} : \mathcal{D} \to 2^G$ is defined, for $d \in \mathcal{D}$ by $\beta_{\mathcal{D}}(d) = \{g \in G : d \sqsubseteq \delta_{\mathcal{D}}(d)\}$.

Pattern concepts are pairs (A,d), $A \subseteq G$, $d \in \mathcal{D}$ such that $\alpha_{\mathcal{D}}(A) = d$ and $A = \beta_{\mathcal{D}}(d)$. d is a pattern intent, and is the common description of all objects in A. When partially ordered by $(A_1,d_1) \leq (A_2,d_2) \iff A_1 \subseteq A_2 (\iff d_2 \sqsubseteq d_1)$, the set of all pattern concepts forms a lattice called the *pattern lattice*.

A nice result in pattern structure establishes that there is a Galois connection between G and $\mathcal{D}^i$ if and only if $(\mathcal{D}^i, \sqcap)$ is a meet semi-lattice.

4.2.2 NextPriorityConcept

4.2.2.1 Predicates for Heterogeneous Data

Our NextPriorityConcept algorithm considers a heterogeneous dataset (G,S) as input where each characteristic $s \in S$ associates to any $a \in G$ either a numerical value, or a binary one, a categorical one, a sequence, a graph, etc. ... i.e. a value depending on the type of s.

Characteristics are given by a family $S = (S^i)_{i \leq d}$, where each S^i contains characteristics on the same type. Different characteristics must be processed separately whereas similar characteristics can be processed together or separately, and some characteristics may not be considered, or considered several times in different groups S_i.

We use monadic predicates to describe characteristics in a generic way. For example, a numerical characteristic $s \in S$ describe any object $a \in G$ by predicates of the form $s(a)$ *is lesser/greater than* c where c is a numerical value. As another example, to a sequence $s \in S$ are associated predicates of the form $s(a)$ *contains x as subsequence"* where x is a sequence. Let p_s be a predicate for a characteristic s and a an object, we note $p_s(a)$ when $s(a)$ verifies p_s. We denote by P the final set of monadic predicates that can be considered as attributes for any $a \in G$, therefore our algorithm computes concepts on $G \times P$, To avoid confusion, we will denote such a concept (A, D) instead of (A, B). Let us notice here that we could have use the term of attributes or features instead of monadic predicate, but we wanted to make a difference between classical input attributes, and our monadic predicates generated by the descriptions.

4.2.2.2 Descriptions and Strategies

We introduce the notions of *description* δ to provide predicates describing a set of objects A, and of *strategy* σ to provide predicates (called *selectors*) refining a concept into smaller concepts.

Predicates describing the objects A are locally computed by a specific treatment for each group of characteristics S^i, and the final description δ is the union of these predicates. Each concept $(A, \delta(A))$ is then composed of a subset of objects A together with a set of predicates $\delta(A)$ describing them. Our NEXTPRIORITYCONCEPT algorithm can be interpreted as a pattern structure on each domain of characteristics S^i. We have an implicit description space $\mathscr{D}^i$ by δ^i for each subset S^i of characteristics, where the description $\delta^i(A)$ of a set $A \subseteq G$ is a direct translation by predicates of its description in $\mathscr{D}^i$. As in pattern structures, the descriptions must verify $\delta^i(A) \sqsubseteq \delta^i(A')$ for $A' \subseteq A$ in order to obtain a lattice. However, unlike classical pattern structures, predicates are not globally computed in a preprocessing step, but locally for each concept.

The algorithm also introduces the notion of *strategy* σ to provide monadic predicates (called *selectors*) to refine the description $\delta(A)$ to a reduced set $A' \subseteq A$ of objects, and $(A', \delta(A'))$ is then a predecessor of $(A, \delta(A))$ in the lattice. Selectors can propose a restriction on the description $\delta(A)$ or on the set A. They are defined according to each group characteristic S^i, and the final strategy σ is the union of the selectors σ^i for each characteristic.

Several strategies are possible to generate predecessors of a concept, going from the naive strategy classically used in FCA that considers all the possible predecessors, to strategies reducing the number of predecessors in order to obtain smaller lattices. Let us observe that selectors are only used for the predecessors generation, they are not kept either in the description or in the final set of predicates. Therefore, choosing or testing several strategies at each iteration in a user-driven pattern discovery approach would be interesting. It is also possible to introduce a filter (or meta-strategy) on the selectors as for example maximal support (number of objects) of a concept, or entropy (according to a class information).

More formally, $\delta(A)$ and $\sigma(A)$ are the union of descriptions and strategies for each group S^i of characteristics where:

A description δ^i is an application $\delta^i : 2^G \rightarrow 2^P$ which defines a set of monadic predicates $\delta^i(A)$ describing the characteristics of S^i for any subset A of G.

A strategy σ^i is an application $\sigma^i : 2^G \rightarrow 2^P$ which defines a set of selectors $\sigma^i(A)$ for characteristics of S^i from which the predecessors of a concept (A, D) are generated.

4.2.2.3 Bordat's Algorithm, Priority Queue and Propagation of Constraints

Our NextPriorityConcept is based on Bordat's algorithm [5], that we also find in Linding's work [13]. This algorithm recursively computes the Hasse diagram of the concept lattice of a context, starting from the bottom concept and by computing at each recursive call the immediate successors of a concept using Bordat's theorem. We extend the dual version of Bordat's theorem stating that the immediate predecessors of a concept (A, D) are the maximal inclusion subsets of the following family on the binary attributes M:

$$\mathscr{F}_{(A,D)} = \{\beta(b) \cap A \: : \: b \in M \setminus D\} \tag{4.1}$$

In our algorithm, the recursion of Bordat's algorithm is replaced by a priority queue using the support of concepts. At each iteration, the concept of maximal support is produced, then its immediate predecessors are computed. Therefore concepts are generated level by level, starting from the top concept, and each concept is generated before its predecessors. Moreover, rather than considering the whole set of possible attributes of $M \setminus D$ to calculate the potential predecessors of a concept (A, D), we consider the selectors $\sigma(A)$ issued from the strategy σ. Then, to generate a lattice, we must ensure that meets are correctly generated. For that, we introduce a constraint propagation mechanism $\mathscr{C}$ that associates a set of selectors $\mathscr{C}[A]$ to process to each concept (A, D) by $\mathscr{C}[A] = C_{\text{residual}} \cup C_{\text{cross}}$ where C_{residual} is the set of residual constraints/selectors issued from the concept that generated (A, D) and C_{cross} is the set of cross constraints/selectors issued from the other predecessors of (A, D). These constraints are initialized by $\mathscr{C}[G] = \emptyset$.

Therefore, we consider the selectors $p \in \sigma(A) \cup \mathscr{C}[A]$, and we compute the predecessors of a concept (A, D) by computing the maximal inclusion subsets of:

$$\mathscr{FD}_{(A,D)} = \{\{a \in A \ : \ p(a)\} \ : \ p \in (\sigma(A) \cup \mathscr{C}[A])\} \tag{4.2}$$

4.2.2.4 Description of the Algorithm

The NEXTPRIORITYCONCEPT algorithm (Algorithm 2) computes the formal context $\langle G, P, I_P \rangle$ and its concepts, where P is the set of monadic predicates describing the characteristics, $I_P = \{(a, p) \ : \ p(a)\}$ is the relation between objects and predicates, and (α_P, β_P) is the associated Galois connection. This algorithm considers as input a dataset $\langle G, S \rangle$ with characteristics organized into groups S^i, a description δ^i and a strategy σ^i for each group, and calls the PREDECESSORS-DESC algorithm (Algorithm 3) that computes the predecessors of a concept (A, D) according to the strategy σ as stated in Eq. (4.2).

Theorem 4.1 *If each description δ^i verifies $\delta^i(A) \sqsubseteq \delta^i(A')$ for $A' \subseteq A$, then the NEXTPRIORITYCONCEPT algorithm computes the concept lattice of $\langle G, P, (\alpha_P, \beta_P) \rangle$ with a run-time in $O(|\mathscr{B}| |G| |P|^2 (c_\sigma + c_\delta))$ (where $\mathscr{B}$ is the number of concepts, c_σ is the cost of the strategy and c_δ is the cost of the description, and a space memory in $O(w |P|^2)$ (where w is the width of the concept lattice).*

Table 4.1: `Digit` context where **c** stands for **c**omposed, **e** for **e**ven, **o** for **o**dd, **p** for **p**rime and **s** for **s**quare

	c	e	o	p	s
0	✓	✓			✓
1			✓		✓
2		✓		✓	
3			✓	✓	
4	✓	✓			✓
5			✓	✓	
6	✓	✓			
7			✓	✓	
8	✓	✓			
9	✓		✓		✓

Algorithm 2: NEXTPRIORITYCONCEPT

Data:

- $\langle G, S \rangle$ a dataset
- $(S^i)_{i \leq d}$ a family of S
- δ a description
- σ a strategy

Output:

- the formal context $\langle G, P, I_P \rangle$
- its concepts (A, D)

```
1 begin
2     /* Priority queue for the concepts */
3     Q ← [] ; /* Q is a priority queue using the support of concepts */
4     Q.push((|G|, (G, δ(G)))) ;                 /* Add the top concept into Q */
5
6     /* Data structure for constraints */
7     𝒞 ← [] ; /* 𝒞 is the descendant constraints map being ∅ by default
          */
8     /* Data structures for predicates */
9     P ← ∅ ;                          /* P is the set of all predicates */
10    I_P ← ∅ ;             /* I_P is the binary relation between G and P */
11    /* Immediate predecessors generation */
12    while Q not empty do
13        (A, D) ← Q.pop() ;       /* Get the concept with highest support */
14        produce (A, D) ;
15        LP ← PREDECESSORS-DESC((A, D), P, I_P, 𝒞, σ, δ) ;
16        /* Update queue */
17        forall (A', D') ∈ LP do
18            if (A', D') ∉ Q then
19                Q.push((|A'|, (A', D'))) ;            /* Add concept into Q */
20        delete 𝒞[A] ;                          /* Remove useless data */
21    return ⟨G, P, I_P⟩
```

Table 4.2: Execution

Step	$\mathscr{C}$	Q	(A, P)	$\|A\|$
0	{0123456789:∅}	[(10,\$0)]	(0123456789,∅)	10
1	{04689:eo, 02468:co, 13579:ce}	[(5,\$1), (5,\$2), (5,\$3)]	(04689,c)	5
2	{02468:co, 13579:ce, 0468:o}	[(5,\$2), (5,\$3), (4,\$4)]	(02468,e)	5
3	{13579:ce, 0468:o}	[(5,\$3), (4,\$4)]	(13579,o)	5
4	{0468:o, 357:e, 9:e}	[(4,\$4), (3,\$5), (1,\$7)]	(0468,ce)	4
5	{357:e, 9:e, 04:o}	[(3,\$5), (2,\$6), (1,\$7)]	(357,op)	3
6	{9:e, 04:o, ∅:ceops}	[(2,\$6), (1,\$7), (0,\$8)]	(04,ces)	2
7	{9:e, ∅:ceops}	[(1,\$7), (0,\$8)]	(9,cos)	1
8	{∅:ceops}	[(0,\$8)]	(∅,ceops)	0

Step $\mathscr{C}$		Q	(A,P)	$|A|$
9 {}		[]		

Table 4.3: Context of the lattice with the maximal support strategy where ($\checkmark$) stands for the relations that are not considered

	c	e	o	p|o	s|ec
0	$\checkmark$	$\checkmark$			$\checkmark$
1		$\checkmark$		($\checkmark$)	
2		$\checkmark$	($\checkmark$)		
3		$\checkmark$	$\checkmark$		
4	$\checkmark$	$\checkmark$			$\checkmark$
5		$\checkmark$	$\checkmark$		
6	$\checkmark$	$\checkmark$			
7		$\checkmark$	$\checkmark$		
8	$\checkmark$	$\checkmark$			
9	$\checkmark$	$\checkmark$	($\checkmark$)		

We consider the formal context `digit` in Table 4.1 as an illustrative example. The maximal support strategy σ_{max} leads to the lattice whose Hasse diagram is displayed in Fig. 4.1, where attributes and objects are indicated in respectively the first and the last concept where they appear. We also indicate the number (using $) and the support (using #) of each concept. The trace execution is in Table 4.2. The classical concept lattice produced without strategy is displayed in Fig. 4.2. The resulting context (G, P, I_P) displayed in Table 4.3 has been constructed using the initial one.

We can observe that the concept lattice with strategy contains only 9 concepts instead of 14. The concepts for attributes **p** and **s** are not generated as immediate predecessors of the top concept since their support is not maximal. However, **p** appears in concept $5, generated as a predecessor of concept $3 equal to $(\{1,3,5,7,9\}, o)$, thus introduced only for the odd digits $\{1,3,5,7,9\}$.

4.2.2.5 The GALACTIC Platform

Such a use of predicates whatever the characteristics allows a generic implementation of our algorithm which, mixed with a system of plugins, makes it possible to envisage an easy integration of new characteristics, new description, new strategies and new meta-strategies. We have developed a code diffusion via a development platform,

---

Algorithm 3: PREDECESSORS-DESC

Data:
- (A, D) a concept
- P the set of predicates
- I_P the binary relation between G and P
- $\mathscr{C}$ the constraints
- σ a strategy (issued from the σ^i)
- δ a description (issued from the δ^i)

Result:

- LP a set of predecessors

```
1  begin
2  │  L ← ∅;
3  │  forall p ∈ (σ(A) ∪ 𝒞[A]) \ D do
4  │  │     /* p is a new "potential" selector to generate a predecessor
      │  │        */
5  │  │     A' ← {a ∈ A : p(a)} ;        /* A' are the objects verifying D+p */
6  │  │     /* Add (A',p) if A' maximum in L and included in A */
7  │  │     if A' ⊆ A then L ← INCLUSION-MAX(L,(A',p));
8  │  N ← {p : (A',p) ∈ L} ;             /* N is the set of new constraints */
9  │  LP ← ∅;
10 │  forall (A',p') ∈ L do
11 │  │     /* Update the selected attributes P and the relation I_P*/
12 │  │     if p' ∈ σ(A) then
13 │  │     │  P ← P ∪ {p'} ;      /* Update the set of selected predicates */
14 │  │     D' ← δ(A') ;           /* D' are the new predicates describing A' */
15 │  │     LP.add((A',D')) ;               /* (A',D') is a new concept */
16 │  │     I_P ← I_P ∪ (A' × D') ;         /* Update the new relation */
17 │  │     /* Compute cross constraints (X) and propagate constraints */
18 │  │     X ← {p'' ∈ N : p''(a) ∀a ∈ A'}
19 │  │     𝒞[A'] ← 𝒞[A'] ∪ 𝒞[A] ∪ N \ X
20 │  return LP
```

---

called **GALACTIC**[1] (see [1] for all icons used in the **GALACTIC** project), for an easy integration of new plugins.

The **GALACTIC** eco-system is organized with (cf Fig. 4.3):

 A core library which implements the NEXTPRIORITYCONCEPT algorithm and a lot of tools for visualizing lattices and reduced contexts in python notebooks;

 A set of characteristic plugins. These plugins define new types of data;

 A set of description plugins. These plugins define new types of descriptions and their associated predicates;

---

[1] https://galactic.univ-lr.fr.

Algorithm 4: INCLUSION-MAX

Data: L a set of potential predecessors represented by their extent, (A,b) a new potential
     predecessor A with its selector predicate b
Result: inclusion maximal subsets of $L+A$

```
1  begin
2  │  add ← true;
3  │  forall (A′,b′) ∈ L do
4  │  │  if A ⊆ A′ then
5  │  │  │  add ← false ;              /* A is not an immediate predecessor */
6  │  │  │  break
7  │  │  else if A′ ⊆ A then
8  │  │  └  L.remove((A′,b)) ;         /* Remove A′ as a possible predecessor */
9  │  if add then L.add(A);                       /* Add (A,b) as a predecessor */
10 │  return L
```

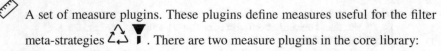

A set of strategy plugins. These plugins define new types of strategies for a given characteristic;

A set of measure plugins. These plugins define measures useful for the filter meta-strategies $\overset{\wedge}{\smile}$ ▼ . There are two measure plugins in the core library:

  The confidence measure

  The support measure

A set of data reader plugins. These plugins allow to read various type of data file. We support:

  boolean reader plugins (compatible with old file formats of FCA);

  heterogenuous reader plugins (*CSV, TOML, INI*);

  complex reader plugins (*JSON, YAML*).

A set of applications using the core library and the different plugins;

A set of localization plugins for translating the different applications.

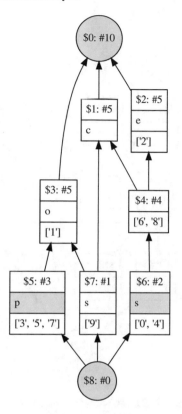

Fig. 4.1: `Digit` sample with greatest support strategy

All plugins can be added *on the fly* to the core engine. We have already developed:

✅ Logical plugins able to use either:

  – the classical exploration in Formal Concept Analysis;
  – the possibility to generate predicates using logical formulae involving a group of several boolean attributes and their negations [16].

🎛 Numerical plugins describing a collection of objects by a convex hull on $\mathbb{R}^n$, i.e. a group of n numerical characteristics. The *Normal* strategy consists in restricting the set of objects using $m \pm \alpha\sigma$ on the principal component (using Principal Component Analysis techniques). α is an input parameter of the strategy.

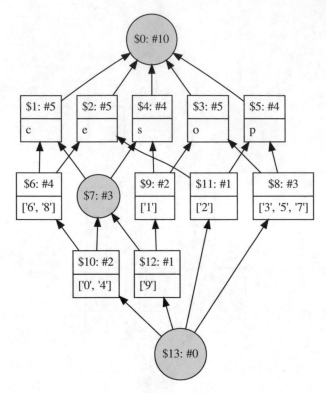

Fig. 4.2: `Digit` sample without strategy

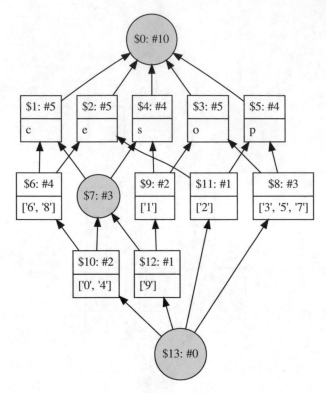 Another numerical *Quantile* strategy plugin restricting the set of objects using a quantile based approach.

Categorized plugins describing a collection of objects having a categorized characteristic by the minimal subset containing their values. The strategy consist in removing a value from this subset to select a smaller set of objects.

String plugins describing a collection of objects having a string characteristic by a set of regular expressions which roughly correspond to the maximal common subchain. The strategy consist in adding a regular expression to select a smaller set of objects.

Chain characteristic and strategy plugins describing a collection of objects having a chain characteristic, or we can say "list". Chains are a collection of successive elements that appear in an ordered list numbered by integers starting from 0.

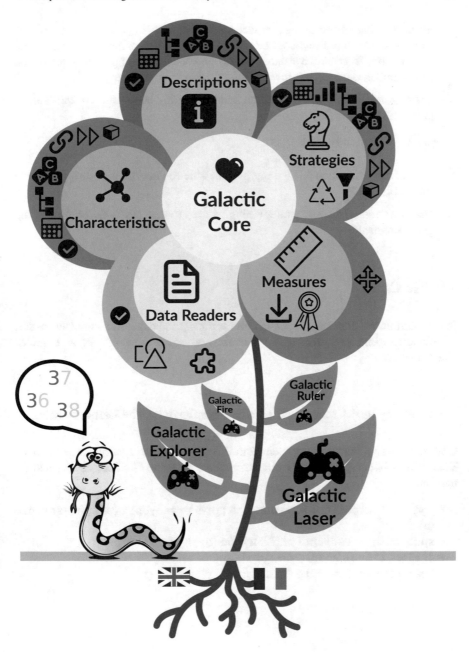

Fig. 4.3: GALACTIC architecture

Sequence plugins describing a collection of objects having a sequence characteristic. A sequence is a set of sorted elements, each element having a comparable
information. There exist a distance notion between these comparable information
(for example, timestamps are valid information for sequences).

 Entropy measure for computing the gain of entropy between a concept and its
predecessors.

And we are currently working on:

 Triadic characteristic and strategy plugins able to generate a description using
triadic predicates.

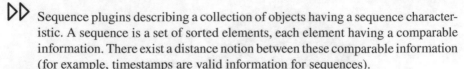 Graph characteristic and strategy plugins able to generate a description using
graph predicates.

4.3 Use Cases

In this section, we illustrate our algorithm for binary and categorical characteristics,
for numerical characteristics, and for chains, strings and sequences as complex
characteristics.

4.3.1 Binary and Categorical Characteristics with the Lenses Dataset

Consider the Lenses dataset issued from the UCI Machine Learning Repository.[2]
This dataset is composed of 24 objects/patients described by four categorical attributes

- age of the patient (G): young (y) ; pre-presbyopic (pp) ; presbyopic
  (p)
- spectacle prescription (P): myope (m) ; hypermetrope (h)
- astigmatic (A): no (n) ; yes (y)
- Tear production rate (T): reduced (r) normal (n)

---

[2] https://archive.ics.uci.edu.

and classified in three classes (C):

- the patient should be fitted with hard contact lenses (h)
- the patient should be fitted with soft contact lenses (s)
- the patient should not be fitted with contact lenses (n).

We consider the modalities of the 4 categorical attributes as binary attributes, i.e. 9 binary attributes.

4.3.1.1 Lenses with the Entropy as Strategy

We consider the classical binary description of a subset A of patients as a set of predicates:

for each of the 9 binary attributes s the classical binary description of a subset A of patients by the set of its attributes:

$$\delta(A) = \{s == x \; : \; s(a) = x \; \forall \text{attribute } s \; \forall a \in A\}$$

We use the entropy as strategy, a supervised strategy using the class information:

$$\sigma_{\text{entropy}}(A) = \{s == b \; : \; b \in \bigcup_{s^i} \delta^i(A) \; : \; H_{\text{class}}(\beta(b)) \text{ minimal}\}$$

The classical FCA approach would be to consider all possible attributes as strategy, and the classical concept lattice contains 110 concepts. With the entropy strategy using the class information and by keeping only the two best entropy measures for the predecessors, we obtain a more compact lattice of 16 concepts displayed in Fig. 4.4.

The 9 concepts $3, $9, $15, $17, $18, $10, $14, $23 and $24 are composed of objects belonging to the same class, and can be interpreted as a clustering of the data into 9 clusters, each concept (A, D) among these 7 concepts corresponding to a cluster (c, D), where c is the class of the objects of A, and meaning that objects having attributes D belong to the class c.

4.3.1.2 Lenses with the Minimal Logic Formulae as Description

When some characteristics are defined on the same domain, our algorithm offers the possibility to process them together. An immediate way to process with a group of characteristics would be to merge the predicates obtained in the individual case, both for the descriptions and the strategies. But it is possible to obtain more relevant predicates by a specific process of a group of characteristics. In Lenses data set, we consider four groups of binary attributes:

- the modes of age of the patient together;
- the modes of spectacle prescription together;
- the modes of astigmatic together;
- the modes of tear production rate together.

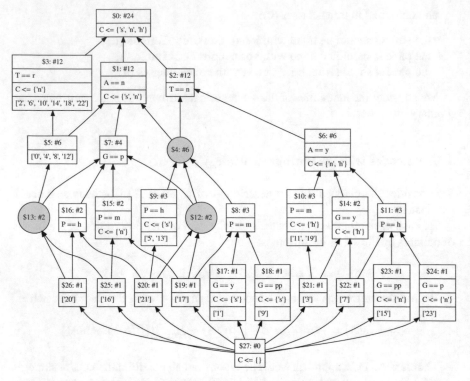

Fig. 4.4: Lenses dataset with the entropy strategy

We produce a finer description for a group of k binary attributes with the introduction of negative attributes, the description $\delta(A)$ is issued from the minimization of the disjunction of clauses:

$$\bigvee_{a \in A} \bigwedge_{j \leq k} \begin{cases} x_j & \text{if } x_j(a) = 1 \\ \overline{x_j} & \text{if } x_j(a) = 0 \end{cases}$$

To compute the minimization of this disjunction of clauses, we use the well-known Quine-McCluskey algorithm (or the method of prime implicants) that minimizes a boolean formulae with a time complexity in $O(3^n \log n)$ [16].

With this description and the entropy strategy, we obtain the lattice displayed in Fig. 4.5 with 40 concepts. We can observe that this concept lattice has more concepts that the concept lattice of Fig. 4.4. This is due to the fact that the introduction of attribute negations tends to generate more concepts in general. However, there are only 6 concepts composed of patients in the same class, instead of 9 in the previous experiment.

4.3.2 Numerical Characteristics with the `Iris` Dataset

Consider the well-known `Iris` dataset issued from the UCI Machine Learning Repository,[3] and composed of 150 objects described by four numerical characteristics:

- `sepal-length`
- `sepal-width`
- `petal-length`
- `petal-width`

and classified in three classes:

- `Setosa`
- `Versicolor`
- `Virginica`

4.3.2.1 `Iris` with the Entropy as Strategy.

We consider the four `petal` and `sepal` characteristics separately, each of these numerical characteristics s provided with a description δ^i for a subset A of objects:

$$\delta^i(A) = \{is \ greater \ than \ \min_{a \in A} s(a), is \ lesser \ than \ \max_{a \in A} s(a)\}$$

We use the entropy as meta-strategy, a supervised strategy using the class characteristics:

$$\sigma_{\text{entropy}}(A) = \{b \in \bigcup_{s^i} \delta^i(A) \ : \ H_{\text{class}}(\beta(b)) \ \text{minimal}\}$$

In order to consider the entropy of a predecessor A' of A, but also the entropy of the remaining set $A \setminus A'$, we defined H by:

$$H = \theta(H_{A'}) + (1 - \theta)H_{A \setminus A'}$$

The more the value of θ increases, the more the number of predessors of A decreases.

If we apply these strategies to the `Iris` data set, we obtain the concept lattice displayed in Fig. 4.6 (with $\theta = \frac{1}{2}$) composed of 13 concepts. The `Setosa` iris are quickly separated according to their two `petal` characteristics (concept $3). Indeed, we can observe on the scatterplot[4] in Fig. 4.7 that this class is clearly separated from the two others using the petal measures which have been chosen by the filtering

---

[3] https://archive.ics.uci.edu.

[4] The Iris dataset scatterplot has been taken from https://commons.wikimedia.org/wiki/File:Iris_dataset_scatterplot.svg.

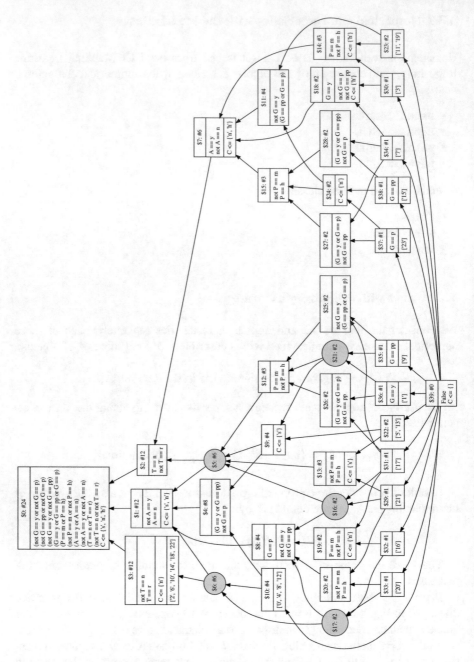

Fig. 4.5: Lenses dataset with the convex hull strategy

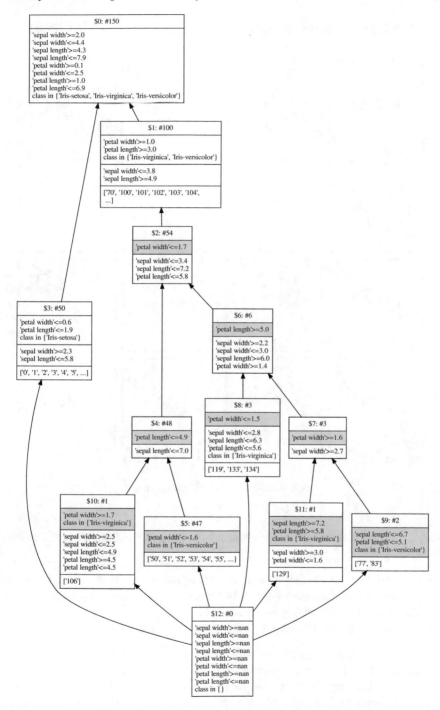

Fig. 4.6: `Iris` dataset with the entropy strategy

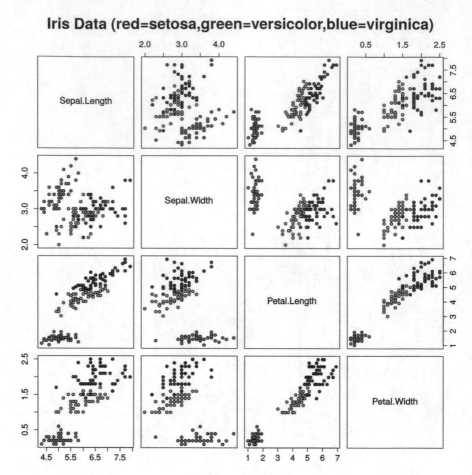

Fig. 4.7: `Iris` dataset scatterplot

strategy as the first selectors. We obtain 4 concepts for classes `Virginica` and `Versicolor`:

- concept $10 and $11 correspond to objects whose class is `Virginica`, with only the two `petal` characteristics used in concept $10, while the `sepal-length` characteristic is introduced in concept $11.
- concept $5 and $9 correspond to objects whose class is `Versicolor`,

Our filtering strategy can also manage all the selectors corresponding to the n best measures. In Fig. 4.6, only the selectors with the **first** best measure were examined. We did the same experiment by varying n from 1 (see Fig. 4.6) to 10. Table 4.4 shows the number of concepts obtained as well as the number of meet irreducible as a function of n. This example shows that it would be interesting to propose a tool for the data scientist where the user could choose or test several strategies for each concept in a interactive way.

Table 4.4: Iris data set using filtering strategy with entropy and keeping the selectors corresponding to the best n measure

n	Concepts	Irreducible
1	13	7
2	42	15
3	275	38
4	981	61
5	2445	75
6	5164	83
7	9639	99
8	19116	126
9	31776	138
10	49241	143

4.3.2.2 Iris with the Convex Hull as Description

We consider groups of numerical characteristics:

- the group of sepal-length and sepal-width together;
- the group of petal-length and petal-width together;

More generally, for a group of k numerical characteristics $s_1, \ldots s_k$, we consider the k-dimensional points $\{(s_j(a))_{j \leq k} : a \in A\}$ for a set A of objects, and their convex hull. The description $\delta^i(A)$ is then composed of predicates describing the borders of the convex hull, and the strategy $\sigma^i(A)$ is a way to reduce the hull. For points in two dimensions, the convex hull is a polygon, and borders and cuts are lines. Clearly, for two sets A and A' of objects such that $A \subseteq A'$, the convex hull of A is included into the convex hull of A', and the intersection of two convex hulls is a convex hull. Therefore $\delta^i(A) \sqsubseteq \delta^i(A')$. For points in two and three dimensions, output-sensitive algorithms are known to compute the convex hull in time $O(n \log n)$. For dimensions d higher than 3, the time for computing the convex hull is $O(n^{\lfloor d/2 \rfloor})$ [6]. This process therefore impacts on the costs c_σ and c_δ.

We have studied in Fig. 4.8 a description by convex hull, and the mixing of three strategies:

- the cuts of the convex hull of sepal-length and sepal-width;
- the cuts of the convex hull of petal-length and petal-width;
- the *categorized* strategy

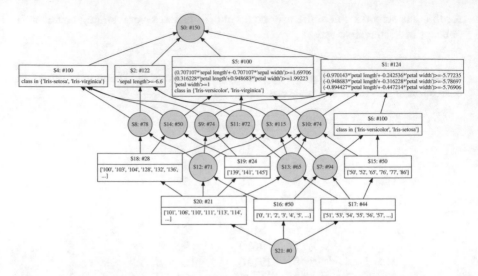

Fig. 4.8: `Iris` dataset with 2 convex hulls in 2 dimensions (aggregating the sepal and petal measures) (limiting the support to 0.81), and the class parameter

The two first strategies were limited to keep new generated predecessors with a minimum support of 0.81. This lead to a concept lattice with 22 concepts:

- class `Virginica` is described in concept $14;
- class `Versicolor` is described in concept $15;
- class `Setosa` is described in concept $16;

4.3.3 Sequential Characteristics with the `Daily-actions` Dataset

We consider a set of sequences, then each object is described by one sequence $s = < (s_i, t_i) >_{i \leq n}$ where t_i is a time information, and s_i is an event defined on an alphabet Σ. The *daily-actions dataset* is a small base of sequences representing daily actions of persons, where persons are member of our L3i laboratory.[5] The daily actions are:

$$\Sigma = \{ \text{Wakeup, Breakfast, Work, Coffee, Lunch,}$$
$$\text{Sports, Dinner, Read, Rest, Sleep, Other} \}$$

Two examples of sequences, where the identifiers correspond to hours:

- $<$(8,Wakeup), (9,Breakfast), (10,Work), (12,Coffee), (13,Lunch), (20,Sports), (22,Dinner), (23,Read)$>$

---

[5] https://l3i.univ-larochelle.fr/.

- <(8,Wakeup), (9,Work), (10,Coffee), (11,Breakfast), (13,Lunch), (14,Sports), (21,Dinner)>

4.3.3.1 Daily-actions with the Common Subsequences

We provide sequences with the following description δ and strategy σ for a subset A of objects and an integer l:

$$\delta^i(A,l) = \{ \; contains \; x \; \text{as subsequence} \; :$$
$$x \text{ is a maximal common subsequence of sequences in } A \text{ and } |X| = l\}$$
$$\sigma^i(A,l) = \{ \; contains \; x \; \text{as subsequence} \; :$$
$$x \text{ is a maximal common subsequence of sequences in } A' \subseteq A \text{ and } |X'| = l\}$$

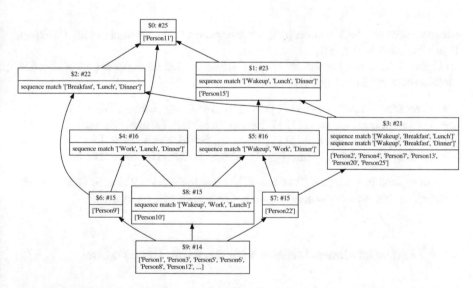

Fig. 4.9: `Daily-actions` dataset using the simple match strategy with length 3 and cardinality 15

We use the `limit-strategy` as meta-strategy, that limits the number of generated predecessors to those of support greater than a given threshold. We obtain the concept lattice displayed in Fig. 4.9 (with $l = 3$) composed of 10 concepts. We observe a difference between concepts \$1, \$2, and concept \$4. The first two concepts correspond to sequences like [Breakfast, Lunch, Dinner] and [Wakeup, Lunch, Dinner], and even in concept \$3 where in addition we get [Wakeup, Breakfast, Lunch], and [Wakeup, Breakfast, Dinner]. While in concepts \$4, \$5, \$8, the element Work appear in the subsequences defining these concepts. We can also observe that the 4

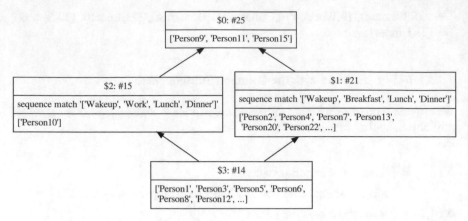

Fig. 4.10: `Daily-actions` dataset using the simple match strategy with length 4 and cardinality 15

subsequences of size 3 in concept $3 are composed of the same elements {Wakeup, Breakfast, Lunch, Dinner}.

Figure 4.10 shows the lattice with $l = 4$. We can see that concept $3 in Fig. 4.9, matches concept $1 in Fig. 4.10:

- [Breakfast, Lunch, Dinner] ⊑ [Wakeup, **Breakfast**, **Lunch**, **Dinner**]
- [Wakeup, Lunch, Dinner] ⊑ [**Wakeup**, Breakfast, **Lunch**, **Dinner**]
- [Wakeup, Breakfast, Lunch] ⊑ [**Wakeup**, **Breakfast**, **Lunch**, Dinner]
- [Wakeup, Breakfast, Dinner] ⊑ [**Wakeup**, **Breakfast**, Lunch, **Dinner**]

And we find that concept $2 of Fig. 4.10 contains Work, [Wakeup, Work, Lunch, Dinner], so that we can distinguish two major groups of subsequences.

4.3.4 Sequential Characteristics with the `Wine City` Dataset

The `Wine City` dataset is based on data provided by the "La cité du vin" museum in Bordeaux, France,[6] gathered from the visits on a period of one year (May 2016 to May 2017). This very special kind of museum is working with a "Visit companion". When visitors arrive at the museum, they receive this little personal device. The `Wine city` uses this device to detect whenever a visitor is close to an animation spot, and automatically play a video, sound or animation. An animation spot of the wine city is called a `module`. As you can see in Fig. 4.10, the museum is a big "open-space" kind of museum, where visitors are free to explore the museum the way they want, without predetermined path. As you may have already understood, the wine city dataset is indeed, just the extractions of logs from the visit companion

---

[6] https://www.laciteduvin.com/en.

itself. By extracting the sequences of activation of each module, we end up with a good idea of what the visit looked like for each visitor of the museum. From now on, we will consider the sequences of activation of modules as trajectories of visitors.

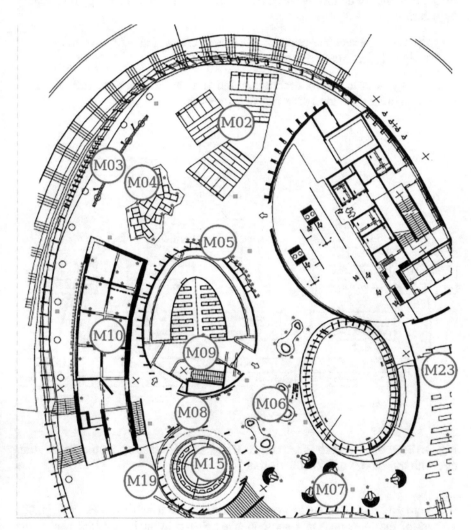

Fig. 4.11: Wine City Context Localization of each module (pics by https://www.laciteduvin.com/fr)

4.3.4.1 `Wine City` Dataset with the Prefix Description and Strategy

The purpose here is to compare each visitor's trajectory in order to find the common trajectories they have made, and study the behaviour of people in open-space visiting situations. To do this, we used the prefix description and strategy on people's trajectories:

- $\delta^i(A) = \{$ *contains x* as prefix : x common prefix of all elements in $A\}$
- $\sigma^i(A) = \{$ *contains* $\delta^i(A) + x'$ as prefix : $x' \in \Sigma\}$

Basically, with the prefix description and strategy, we compare suffixes of sequences in order to build a prefix tree of sequences.

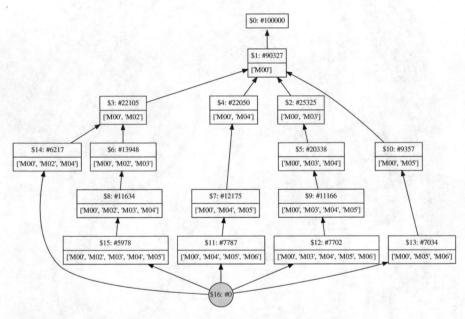

Fig. 4.12: `Prefix match strategy: Wine City dataset` Prefix match strategy with trajectories of 100.000 visitors

Figure 4.12 took 3 minutes to generate and is made from trajectories of 100.000 visitors of the museum. Only subsequences shared by at least 5.000 people are displayed in the figure. Nevertheless, even though we have a small portion of what GALACTIC is capable of, we clearly see the popular starts among visitors. M00 will not appear in the map of the museum as it represents the start of the visit. In the figure, we can see that M02, M04 and M03 seem to be popular starts. An interesting thing to see is that when people "skip" modules (for instance, sequences starting with M04 won't go back to M03 or M02, indicating that people are unlikely to go back to the module they missed). This kind of strategy is a simple way to represent

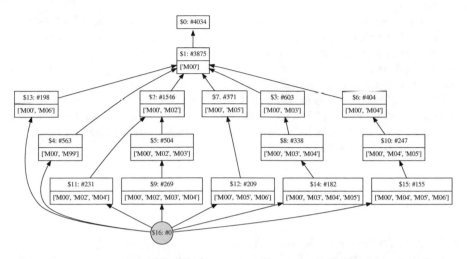

Fig. 4.13: **Prefix match strategy: Wine City dataset** Prefix match strategy with trajectories of young people (Less that 12 years) where M99 refers to the end of the visit

the behaviour of people and is well suited for all kinds of trajectory datasets (after a sequencing process) (Figs. 4.12 and 4.13).

4.4 Conclusion

In this paper, we described our NEXTPRIORITYCONCEPT algorithm constructing formal concepts for heterogeneous and complex data. Inspired from pattern structures and based on Bordat's algorithm, this algorithm considers heterogeneous data as input with a description mechanism and a predecessor generation strategy adapted to each kind of data, and generically described by monadic predicates. The originality of this algorithm is based on the introduction of descriptions and strategies (that locally compute selectors), the use of the priority queue (to ensure that each concept is generated before its predecessors), and the propagation of constraints (to ensure that meets will be computed). The monadic predicates of the descriptions are locally computed, and well suited to the data, lattices are often smaller, with more relevant concepts with the use of specific exploration strategies.

We presented our development platform, called GALACTIC, composed of a generic implementation of the NEXTPRIORITYCONCEPT algorithm and a system of plugins for an easy integration of new characteristics, new descriptions, new strategies and new meta-strategies. Then we presented some use cases and their implemented descriptions and strategies for classical boolean, numerical and categorical characteristics, and for complex data that are strings, chains and sequences.

Future work will be devoted to the implementation of new plugins for intervals, triadic data, graphs . . . Mining many heterogeneous and real datasets would allows us to provide strategies of exploration adapted to each dataset. We also plan an extension of the NEXTPRIORITYCONCEPT algorithm for a generic minimal generator and rule computation. Indeed, the algorithm we have developed transforms the predicates of the initial set of attributes into a new set where each element is derived from a combination of elements from the initial one. We need to study the impacts of this transformation on the rule bases induced by the concept lattice and in particular on the canonical direct basis.

References

1. GALACTIC image credits. https://galactic.univ-lr.fr/credits.html
2. Barbut, M., Monjardet, B.: Ordres et classifications : Algèbre et combinatoire. Hachette, Paris (1970), 2 tomes
3. Bertet, K., Demko, Ch., Viaud, J.F., Guérin, C.: Lattices, closures systems and implication bases: A survey of structural aspects and algorithms. Theoretical Computer Science **743**, 93–109 (2018)
4. Birkhoff, G.: Lattice theory, vol. 25. American Mathematical Soc. (1940)
5. Bordat, J.P.: Calcul pratique du treillis de Galois d'une correspondance. Mathématiques et Sciences humaines **96**, 31–47 (1986)
6. Chazelle, B.: An optimal convex hull algorithm in any fixed dimension. Discrete and Computational Geometry **10**(1), 377–409 (1993)
7. Demko, Ch., Bertet, K., Faucher, C., Viaud, J.F., Kuznetsov, S.O.: NEXTPRIORITYCONCEPT: A new and generic algorithm computing concepts from complex and heterogeneous data. Theoretical Computer Science (Accepted)
8. Ferré, S.: Reconciling Expressivity and Usability in Information Access - From Filesystems to the Semantic Web. Habilitation, University of Rennes 1, France (november 2014)
9. Ferré, S., Ridoux, O.: A logical generalization of formal concept analysis. vol. 1867, pp. 371–384 (Mar 2000)
10. Ganter, B., Kuznetsov, S.: Pattern structures and their projections. In: LNCS of International Conference on Conceptual Structures (ICCS'01). pp. 129–142 (2001)
11. Ganter, B., Wille, R.: Formal Concept Analysis, Mathematical foundations. Springer Verlag, Berlin (1999)
12. Kaytoue, M.: Contributions to Pattern Discovery. Habilitation, University of Lyon, France (february 2020)
13. Linding, C.: Fast concept analysis. In: Working with Conceptual Structures-Contributions to ICC. pp. 235–248 (2002)
14. Pei, J., Han, J., Mortazavi-Asl, B., Pinto, H., Chen, Q., Dayal, U., Hsu, M.C.: Prefixspan: Mining sequential patterns efficiently by prefix-projected pattern growth. In: icccn. p. 0215. IEEE (2001)

15. Prado, A., Jeudy, B., Fromont, E., Diot, F.: Mining Spatiotemporal Patterns in Dynamic Plane Graphs. Intelligent Data Analysis **17**(1), 71–92 (2013)
16. Quine, W.V.O.: The problem of simplifying truth functions. The American Mathematical **59**(8), 521–531 (1952)
17. Wille, R.: Restructuring lattice theory : an approach based on hierarchies of concepts. Ordered scts pp. 445–470 (1982), i. Rival (ed.), Dordrecht-Boston, Reidel.
18. Yan, X., Han, J.: gspan: Graph-based substructure pattern mining. In: Proceedings of the 2002 IEEE International Conference on Data Mining (ICDM 2002). pp. 721–724. IEEE Computer Society, Maebashi City, Japan (December 2002)
19. Yan, X., Han, J., Afshar, R.: Clospan: Mining: Closed sequential patterns in large datasets. In: Proceedings of the 2003 SIAM international conference on data mining. pp. 166–177. SIAM (2003)

Chapter 5
Dealing with Large Volumes of Complex Relational Data Using RCA

Agnès Braud, Xavier Dolques, Alain Gutierrez, Marianne Huchard, Priscilla Keip, Florence Le Ber, Pierre Martin, Cristina Nica, and Pierre Silvie

5.1 Introduction

Many data are inherently relational, and their relations can be complex, numerous, fuzzy and sometimes cyclic. Multi-relational datasets are based on a schema (data model), where entities (objects) of several categories are described by characteristics (attributes, fields) and where relations link objects from two categories (possibly from the same one). Several approaches have been implemented to explore such data [17]. Relational Concept Analysis (RCA), based on Formal Concept Analysis (FCA), has been specifically designed for this task: it builds a classification (a lattice of formal concepts) for each category of objects contained in a dataset, and allows to obtain implication rules including relations between objects [28, 45].

A. Braud • X. Dolques • F. Le Ber
ICube UMR 7537, Université de Strasbourg, CNRS, ENGEES, Strasbourg, France
e-mail: agnes.braud@unistra.fr; dolques@unistra.fr; florence.leber@engees.unistra.fr

A. Gutierrez • M. Huchard (✉)
LIRMM, Univ Montpellier, CNRS, Montpellier, France
e-mail: alain.gutierrez@lirmm.fr; marianne.huchard@lirmm.fr

P. Keip • P. Martin
CIRAD, UPR AIDA, Montpellier, France

AIDA, Univ Montpellier, CIRAD, Montpellier, France
e-mail: priscilla.keip@cirad.fr; pierre.martin@cirad.fr

C. Nica
Nicolae Titulescu University of Bucharest, Bucharest, Romania
e-mail: cristina.nica87@yahoo.com

P. Silvie
IRD, UMR IPME, Montpellier, France

CIRAD, UPR AIDA, Montpellier, France

AIDA, Univ Montpellier, CIRAD, Montpellier, France
e-mail: pierre.silvie@cirad.fr

RCA, as FCA, comes with a major challenge, linked to the fact that dealing with large and complex data produces huge and complex results. Many methods have been proposed to reduce the lattice size, either by reducing the original data (e.g. by granular reduction [55]) or by projection [48], or by reducing the number of concepts to be built (e.g., by thresholding [49]), or by using AOC-posets [15].

Another approach is to help the user to navigate the results, e.g. by focusing on specific subsets of concepts, based on interestingness measures [6, 11], or by using local views and computation on-demand [16, 20]. Regarding RCA, the issue is also to navigate a family of lattices, each concept of a lattice being possibly linked to several concepts of other lattices.

RCA has been applied to multi-relational datasets from various domains, e.g. for the fuzzy semantic annotation of web resources [13], or for the analysis and reengineering of software models [14] and semantic wikis [47]. In previous works, we have applied RCA for exploring hydroecological [15, 38] and agricultural data [29], the two domains considered in the following.

In this paper, we experiment the application of RCA on complex environmental datasets coming from the real world and built under guidance of domain experts. The two application domains are biopesticides and antimicrobial products made from plants (KNOMANA project) and the monitoring of the ecological quality of waterbodies (FRESQUEAU project). In the context of the environmental domains we deal with, the studied datasets can be considered large volume data with regard to the type of data and data collection. In KNOMANA, data are manually collected or revised by experts in scientific publications. The publications are of different types and there is a cross-check in different sources, and a cleaning of information to ensure the data quality. In FRESQUEAU, data are manually collected and manipulated by field biologists in rivers, which differs from data collection from sensors. We show the scope of the RCA process in terms of quantitative opportunities and limits on our datasets, by comparing different algorithms. We also describe an application of RCA to the extraction of graphs from temporal data: the issue is to link sequences of physico-chemical parameter values with bio-indicator values used for assessing the quality of waterbodies. This temporal data pattern extraction shows how we can concretely help domain experts.

As said before, these two datasets have already been studied [15, 29, 38]. In this paper, we propose a synthesis of observations made during these earlier studies, with enhanced datasets, taking into account more information or applied to the whole initial data rather than to an excerpt. We also compare the algorithms with a same metric set on both datasets.

Section 5.2 presents RCA principles while Sect. 5.3 compares RCA with the related work. Section 5.4 introduces the two complex environmental datasets, and compares the results obtained on these datasets by a few algorithms. Section 5.5 describes the variant of RCA used for analysing sequential datasets. Besides, it shows how summarizing interrelated concepts by a graph can help the analysis of the RCA results. Section 5.6 discusses the results and draws up some perspectives.

5.2 Background

Formal Concept Analysis (FCA) has several dimensions, including being a knowledge engineering method based on lattice theory [25]. In its simplest form, FCA deals with datasets formalized into *formal contexts* comprising objects described by attributes (object-attribute contexts). Attributes in formal contexts are sometimes refered as Boolean attributes. For example, the top of Table 5.1 shows four formal contexts: `Biopesticide` describes the toxicity of six biopesticides (from `p1` to `p6`) using two attributes (`toxic`, `nonToxic`); `Bioaggressor` informs on the type of six bioaggressors (from `a1` to `a6`) using two attributes (`worm`, `rodent`); `ProtectedSystem` presents six biological systems to be protected (from `s1` to `s6`) using four attributes (`seed`, `cerealSeed`, `cucurbitSeed`, `leaf`); `Country` localizes four countries (from `c1` to `c4`) in two regions using two attributes (`west`, `east`). A formal context may have a specific shape: it may *partition* the objects with mutually exclusive attributes; it may be *diagonal* if it has the same number of objects and attributes, and each object is described by exactly one attribute (the relation corresponds to a 1–1 mapping).

Biopesticide	Toxic	NonToxic
p1		×
p2		×
p3		×
p4		×
p5	×	
p6		×

Bioaggressor	Worm	Rodent
a1	×	
a2	×	
a3	×	
a4	×	
a5		×
a6		×

ProtectedSystem	Seed	cerealSeed	CurcubitSeed	Leaf
s1	×		×	
s2	×		×	
s3	×	×		
s4	×	×		
s5				×
s6				×

Country	West	East
c1	×	
c2	×	
c3		×
c4		×

Treats	a1	a2	a3	a4	a5	a6
p1						
p2		×				
p3			×			
p4	×	×				
p5				×		
p6						×

Attacks	s1	s2	s3	s4	s5	s6
a1	×					
a2		×				
a3			×			
a4				×		
a5					×	
a6						×

isHostedIn	c1	c2	c3	c4
a1		×		
a2				×
a3	×			
a4	×			
a5		×		
a6				×

Table 5.1: Relational Context Family. Top: the formal contexts (object-attribute contexts) `Biopesticide`, `Bioaggressor`, `ProtectedSystem`, `Country`. Bottom: the relational contexts (object-object contexts): `treats`, `attacks`, `isHostedIn`

FCA highlights hierarchies of concepts, each concept being composed of a maximal group of objects (extent) and the maximal group of attributes they share (intent). Since only objects `s3` and `s4` share attributes `seed` and `cerealSeed`, `Concept_ProtectedSystem_2= ({s3, s4}, {seed, cerealSeed})` is a concept. For similar reasons, `Concept_ProtectedSystem_4= ({s1, s2, s3, s4},` `{seed})` is another concept. The set of all concepts provided with inclusion between concept extents (from bottom to top) forms a lattice (the concept lattice). `Concept_ProtectedSystem_2` is a subconcept of `Concept_Protected System_4` in this lattice, as the extent of the former is included in the extent of the latter. Figure 5.1 shows the concept lattices associated with `Biopesticide`, `Bioaggressor`, `ProtectedSystem`, and `Country`. In this representation of lattices, the attributes (resp. objects) are written only in the highest (resp. lowest) concept where they appear (their introducer concept) and are inherited top to bottom (resp. bottom to top). For instance, `Concept_Biopesticide_2` groups non toxic biopesticides p1, p2, p3, and p4, `Concept_Bioaggressor_2` groups worms a1, a2, a3, and a4, `Concept_ProtectedSystem_4` groups seeds s1, s2, s3 and s4, and `Concept_Country_1` groups western countries c1 and c2.

FCA and all its extensions are well-founded mathematical frameworks thanks to lattice theory, delivering to experts explainable results on which they can base their decisions. FCA is the reference for building exact hierarchies of object/attribute structures, attributes being possibly complex descriptions, and has no competitor for that feature. Uta Priss notes that "the basic FCA structures have been rediscovered over and over by different researchers and in different settings." [46], emphasizing their fundamental aspect. FCA has also a central position as a swiss knife in knowledge engineering and discovery, as the conceptual structures intrinsically contain the search space for rules of different kinds, traceable knowledge patterns and hierarchical structures [43]. FCA is human centered, suitable for interactive and incremental analyses, with visual presentation of extracted patterns. Besides, FCA extensions enable to deal with complex information: numbers, sequences, graphs, temporal data, etc. without converting datasets into simplified formats.

RCA extends the purpose of FCA to relational data, conforming to a conceptual model, e.g. a UML model. We follow up the example with data conforming to the UML model shown in Fig. 5.2: biopesticides treat bioaggressors that attack protected systems; bioaggressors are hosted in countries. This UML model thus structures the dataset into object categories (here biopesticides, bioaggressors, protected systems and countries), objects still being described by attributes. Relationships connect objects of different (or the same) categories: here `treats` connects biopesticides to bioaggressors; `attacks` connects bioaggressors to protected systems; `isHostedIn` connects bioaggressors to countries. These relations are shown at the bottom of Table 5.1.

The UML model and its instantiation are formalized as a Relational Context Family (RCF). An RCF is a pair $(\mathcal{K}, \mathcal{R})$, where $\mathcal{K}$ is a set of object-attribute contexts (formal contexts) and $\mathcal{R}$ is a set of object-object contexts (relational contexts or relations). $\mathcal{K}$ contains n object-attribute contexts $K_i = (G_i, M_i, I_i)$, $i \in \{1, \ldots, n\}$ (*formal contexts*). $\mathcal{R}$ contains m object-object contexts $R_j = (G_k, G_l, r_j)$, $j \in \{1, \ldots, m\}$ (*relational contexts*). $r_j \subseteq G_k \times G_l$ is a binary relation with $k, l \in \{1, \ldots, n\}$. $G_k = dom(r_j)$ is the domain of the relation, and $G_l = ran(r_j)$ is the range of the relation.

The RCA process starts by applying FCA first on each object-attribute context of an RCF. This results in the concept lattices presented in Fig. 5.1.

In the following steps, RCA relies on the construction of particular attributes, called *relational attributes*. These attributes express the relationships an object of

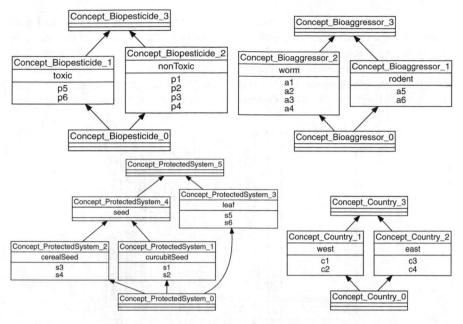

Fig. 5.1: Concept lattices associated with the four formal contexts Biopesticide, Bioaggressor, ProtectedSystem, Country from left to right and top to bottom

Fig. 5.2: Biopesticides treat bioaggressors that attack protected systems. The bioaggressors are hosted in countries

one category has with a concept extent (which is a group of objects of a given category). For example, based on Concept_ProtectedSystem_2 which groups cereal seeds s3 and s4, the relational attribute $\exists attack$(Concept_Protected-System_2), meaning "attack at least one cereal seed", can be formed. This attribute is true for worms a3, a4. This is formalized as follows. A relational attribute $\exists r_j(C)$, where $\exists$ is the existential quantifier, $C = (X,Y)$ is a concept, and $X \subseteq ran(r_j)$, is owned by an object $g \in dom(r_j)$ if $r_j(g) \cap X \neq \emptyset$. Other quantifiers are defined in [9, 28]. In particular, percentage quantifiers are introduced to take into account incomplete, noisy data or approximate satisfaction of a property.

The *relational scaling mechanism* is used to implement the additional description of objects by relational attributes. It maps every relation r_j into a set of *relational attributes* that extend the object-attribute context describing the objects of $dom(r_j)$. This operation is called the relational extension of a context. Table 5.2 shows the relational extension of Bioaggressor at step 1. The first two columns show the original attributes. The next six columns are the relational attributes formed with ∃ quantifier, `attacks` relation, and the concepts of `ProtectedSystem` lattice of step 0. The next four columns are the relational attributes formed with ∃ quantifier, `isHostedIn` relation, and the concepts of `Country` lattice of step 0. From this table, worms a3, a4 own relational attributes ∃*attacks*(`Concept_ProtectedSystem_2`) (cereal seeds) and ∃*isHostedIn*(`Concept_Country_1`) (western countries).

Bioaggressor	Worm	Rodent	∃ attacks(Cpt_ProtectedSystem_3)	∃ attacks(Cpt_ProtectedSystem_1)	∃ attacks(Cpt_ProtectedSystem_0)	∃ attacks(Cpt_ProtectedSystem_2)	∃ attacks(Cpt_ProtectedSystem_4)	∃ attacks(Cpt_ProtectedSystem_5)	∃ isHostedIn(Cpt_Country_2)	∃ isHostedIn(Cpt_Country_0)	∃ isHostedIn(Cpt_Country_1)	∃ isHostedIn(Cpt_Country_3)
a1	×				×				×	×		×
a2	×				×				×	×		×
a3	×					×	×	×		×	×	×
a4	×					×	×	×		×	×	×
a5		×	×						×	×		×
a6		×	×						×	×		×

Table 5.2: Relational extension of `Bioaggressor` at step 1, with relational attributes built on lattices of step 0 (`Cpt` stands for `Concept`)

The application of FCA to all the extended contexts refines the original concept lattices. Figure 5.3 shows the `Bioaggressor` concept lattice at step 1, as a refinement of step 0 (Fig. 5.1). Three concepts are added, in particular `Concept_Bioaggressor_4` which groups a1 and a2, that are worms that attack cucurbit seeds and `Concept_Bioaggressor_5` which groups a3 and a4, that are worms hosted in western countries and attack cereal seeds. These two concepts emerged thanks to the addition of relational concepts and divide the group of worms (`Concept_Bioaggressor_2`).

The complete process operates through successive steps. Each step consists in applying FCA on each object-attribute context extended by the relational attributes created using the concepts from the previous step. This results in a family of concept lattices.

Concept formation propagates from one object category to a neighbouring object category along the relations, refining the concept lattices at each step with concept completion, or new concept addition. To continue on our example, Fig. 5.4 represents the relations and objects in the form of a graph (top, left-hand side); it highlights how groups of non toxic biopesticides, worms, cereal seeds and west-

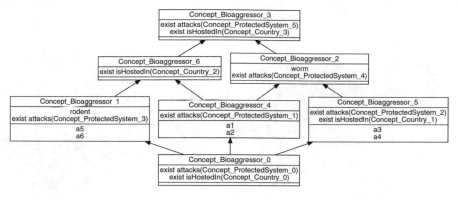

Fig. 5.3: **Bioaggressor** concept lattice at step 1, refining the concept lattice of step 0 (Fig. 5.1)

ern countries are created at step 0 (blue ellipses). Then, at step 1, one can observe the group of worms from western countries attacking cereal seeds (green ellipse). Then at step 2, the group of non toxic biopesticides allowing to treat them is created (red ellipse). This information appears in the lattice of Fig. 5.5, where Concept_Biopesticide_7 groups p2 and p3, that are non toxic biopesticides that treat worms hosted in western countries and attacking cereal seeds (through relational attribute $\exists treats($Concept_Bioaggressor_5$)$).

The RCA process stops when a fixpoint is reached, i.e. when the families of lattices of two consecutive steps are isomorphic and the extended object-attribute contexts are unchanged. The UML model may contain directed cycles, without risk of divergence of the process, as rows (objects) are unchanged, only new columns can be added at each step (with possibly new concepts appearing) and the concept number in each lattice is bounded by $2^{min(|O|,|A|)}$ where O is the object set and A is the attribute set.

5.3 Related Work

Formal Concept Analysis handles multi-relational data through several perspectives. Some approaches extract and classify graph patterns that connect objects or object groups [24, 34, 44]. Relational data have also been dealt with logical concept analysis [23]. Besides, K. E. Wolff has introduced the Relational Semantic Systems: the data model is represented through a conceptual graph, while the relational knowledge is represented through *object traces* and *relation concept traces* in *trace diagrams* [53]. Tuples of Boolean factors are extracted from various tables thanks to an extended version of the Boolean Factor Analysis [31]. An n-ary relation may be in many concrete cases considered as an aggregation of several relations of lower arity. Thus FCA also has been generalized to Triadic Concept Analysis, that considers a ternary

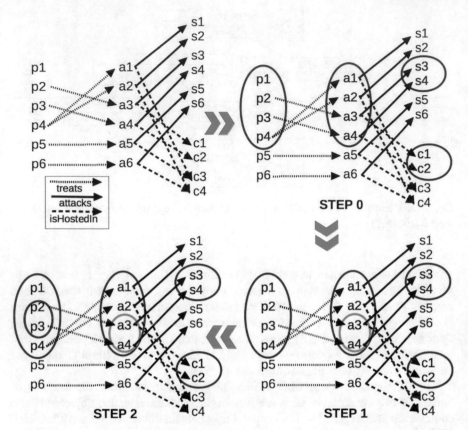

Fig. 5.4: A diagrammatic view on the relational context family with focuses on some concept and sub-concept extents formed at step 0 (top right), step 1 (down right) and step 2 (down left)

relation including objects, attributes and conditions [33]. This yields triadic concepts that are organised in a complete trilattice. This framework has been generalized to n-adic contexts (n-ary relations) in Polyadic Concept Analysis [52].

In [30], a Galois connection (and the derived concept lattice) is introduced to query sets of objects connected by relations. Only existential queries are expressed and there is no iteration. Graph-FCA [21](G-FCA) proposes to consider knowledge graphs based on n-ary relationships as formal contexts. The intent of a G-FCA concept is a projected graph pattern and the extent is an object relation.

Pattern structures [24] can also be used to deal with relational data, for example temporal data. Authors in [10] propose to use pattern structures to build a concept lattice on complex sequential data about care trajectories. The pattern structure is $(P, (S, \sqcap), \delta)$, where P is the set of patients, S is a set of sequences and their sub-sequences, and $\sqcap$ is the set intersection. Each patient of P is described by a sequence (and its sub-sequences) through δ relation. This approach is deepened in [12], where

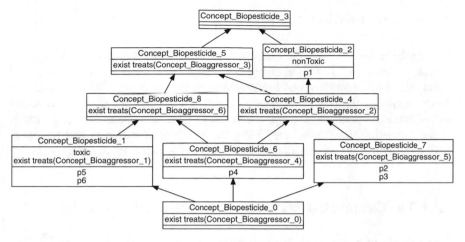

Fig. 5.5: **Biopesticide** concept lattice at step 2 (final step), refining the concept lattice of step 1 (Fig. 5.3)

object descriptions are organised into a semi-lattice of closed sets of closed subsequences. A similar approach is used for analysing demographic sequences in [26].

Compared to these approaches, RCA benefits from several features. Its derivation from the binary framework makes its results more easy to understand than new diagrams introduced in Relational Semantic Systems. It is relevant for incremental data exploration tasks, as it iterates on knowledge construction, showing the progress in concept construction, contrarily to Boolean factor analysis, pattern structures or Graph-FCA. Compared to the other approaches, it provides several operators to take into account incomplete or noisy data. It has been the subject of research on assisting domain experts in the parametrization and exploration [42].

Several papers push the limits of FCA and show effects of application of FCA or RCA in computation time and conceptual structure size on huge or complex datasets. In [54], a huge Museum collection dataset is analysed and made navigable with FCA, showing the efficacy of the recent algorithms. In [37], RCA is applied to UML class model reengineering, with an underlying circular data schema provoking the construction of large amounts of concepts. Such experiments show that (1) FCA can be applied to huge datasets, (2) RCA, that iterates on FCA, is risky in the presence of cycles and has to be handled with care.

In this paper, we focus on a particular kind of datasets, namely in the environmental domain (with observations, plants, animals, etc.), having in mind that they may present some similarities (in the form of data and the form of querying and exploration needs) and that we should learn some lessons when applying RCA for that specific domain.

5.4 RCA for Environmental Data

In this section, we introduce our datasets, the KNOMANA dataset, and the FRESQUEAU dataset in Sect. 5.4.1. In the following Sect. 5.4.2, we analyse the performances of the current RCA algorithms and implementations on two excerpts of these datasets. After describing the UML model of each excerpt, we give the dimensions of the corresponding context family, the computation time of various RCA algorithms when processing these data, and finally the numbers of concepts and relational attributes of the final lattices. Last Sect. 5.4.3 is a discussion about these results.

5.4.1 Two Complex Datasets from the Environmental Domain

5.4.1.1 Pesticidal and Antimicrobial Data

The excessive use of pesticides and antibiotics in agriculture compromises their therapeutic effectiveness and is a threat to human, animal and environmental health [41]. One alternative consists in using natural plant based products. For African farmers, preparing such products using some of the local plants is a challenge. Unfortunately, knowledge on plant use in agriculture is scattered. To support knowledge exchange, description of plants used in Africa was extracted from the scientific literature and collected in a knowledge base called KNOMANA [35]. In KNOMANA, each use of plant is described using 72 data types, among which the protecting plant, the targeted organism (insect, disease, virus, etc.), and the protected system (agricultural crop, animal or human being). In October 2019, KNOMANA gathered 40.800 plant use descriptions for plant, animal, and human health from 410 documents, dated between 1957 and 2019. These uses consider 523 plant protection species, 127 targeted organism species, and 28 protected organism species. In the following (see Sect. 5.4.2) we will explore an excerpt from this database.[1]

5.4.1.2 Water Data

The assessment of aquatic ecosystems, as required by the Water Framework Directive [51], relies on monitoring, which generates large volumes of heterogeneous data from multiple sources at different temporal scales. Actually, when assessing the water quality of watercourses, hydroecologists measure both biological and physico-chemical parameters. In metropolitan France, assessment is done on a network of 1781 sampling sites, called stations. Each station is described by biological data, e.g. the number of individuals for each taxon (animal or plant), and by physico-chemical data, e.g. chemical oxygen demand (denoted DCO), ammonium (denoted NH4), temperature (denoted T), suspended organic matter (denoted MES), etc. Taxons are

---

[1] https://dataverse.cirad.fr/dataverse.xhtml?alias=knomana.

themselves described by qualitative characteristics, called traits. Based on biological data, biological indicators are computed, e.g. the IBGN ("indice biologique global normalisé") that summarizes information from macro-invertebrate samples into a rating [3], or the IBD ("indice biologique diatomées") that summarizes information from micro-alga samples [2]. Stations are also described by physical and contextual characteristics (e.g. they belong to a waterbody). The assessment varies on time: major physico-chemical parameters are analysed 12 times a year, and minor elements four or six times a year; biological sampling is achieved once a year or once every two years.

Data collected from the 1781 sampling sites from 2007 to 2013 have been recorded into a PostgreSQL/PostGIS database that was designed during the ANR 11 MONU 14 FRESQUEAU project.[2] In the following, we will explore two datasets from this database.

- The first dataset focuses on the annual descriptions of the sampling sites from Jan. 2017 to Nov. 2013: each pair (site, year) is described by the annual average measures of physico-chemical parameters, by taxon lists and by geographical parameters (see Sect. 5.4.2.2).
- The second dataset focuses on the temporal dimension of the data: indeed, each sampling site can be described by a sequence of time stamped physico-chemical parameter measures and time stamped biological indicators (see Sect. 5.5).

All these data are public data, and are freely available on the Naiades (Eau France) website.[3]

5.4.2 Experimenting RCA Algorithms

In this section, we assess the ability and limits of RCA and of its current implementation to analyse datasets from the FRESQUEAU and KNOMANA databases.

Both databases can be used in a variety of analyses. To determine the limits of the current RCA implementations, we selected two datasets with representative UML models. These models were encoded into relational context families. Tables 5.3 (Sect. 5.4.2.1) and 5.7 (Sect. 5.4.2.2) describe the formal contexts (object number, attribute number, density) and the relational contexts for each dataset. For the later, only density is indicated as the number of rows and columns results from the source and range formal context object number. The density of a formal context (resp. relational context) is given by the size of the relation (pair number) divided by the object number multiplied by the Boolean attribute number (resp. the source object number multiplied by the range object number).

As the (Boolean) attribute number (column number) corresponds to a scaling of the original (quantitative or multi-valued) attributes, we also indicate these two numbers.

---

[2] http://dataqual.engees.unistra.fr/fresqueau_presentation_gb.

[3] http://www.naiades.eaufrance.fr/acces-donnees.

Then the following conceptual structures were built using the ∃ quantifier: the concept lattice (addIntent/addExtent algorithm [36]), the AOC-poset (Ceres algorithms [32], Pluton [8], and Hermes [7]), and the Iceberg lattice (Titanic algorithm [50], with minimal support 10%, 30% and 40%). Result tables 4, 5, 6 (Sect. 5.4.2.1) and 8, 9, 10 (Sect. 5.4.2.2) present metrics on the running time, step number, concept number and relational attribute number, and whether computing the structure failed for each dataset. We chose the ∃ quantifier, as it generates the largest number of concepts in the worst case, this being the most constraining [9]. Experiments are realized using a laptop with a 4 core Intel i7 2.70 GHz processor.

5.4.2.1 Experiments on KNOMANA Dataset

Figure 5.6 shows the UML model, without cycle, chosen for experimenting RCA algorithms on KNOMANA project. In this model, a Document is described by various multi-valued attributes. A document *owns* a piece of Knowledge with a certain quality. This piece of knowledge *describes* a form of HealthProtection which: *protects* a ProtectedSystem *composed of* ProtectedOrganisms; *targets* a TargetedOrganism; *uses* a Biopesticide *made from* a UsedPlant from which a technician *extracts* PlantParts.

Table 5.3 shows the dimensions of the RCF for the considered excerpt of KNOMANA knowledge base. This RCF is composed of 9 formal contexts and 8 relational contexts. The longest path in the UML model graph is made of 5 edges (from Document to PlantPart). The largest formal contexts are HealthProtection (more than 10000 objects), Biopesticide (more than 5000 objects), UsedPlant (about 4000 objects), and Document (about 3500 objects). Furthermore, Table 5.3 (and this can also be observed in Table 5.3 for the FRESQUEAU dataset) shows that the number of objects (rows in formal contexts) is higher than the number of real objects, because an object is described in various situations (e.g. *Lantana camara* may be described in *n* different documents, leading to *n* occurrences of *Lantana camara* (implicitly observations) in the formal context UsedPlant). Densities are most of the time low (e.g. HealthProtection) to very low (e.g. for *uses*). When the number of objects and attributes are equal, the context may be more complex than a diagonal.

Table 5.4 shows the step number and the running time. The RCA process converged at step 7 (the 7th step being to confirm that the fixpoint is reached). The computing time to construct the concept lattice (FCA) was about 2 minutes. The one for AOC-poset varied between 1 and 8 minutes according to the adopted algorithm, CERES being the most efficient. Iceberg lattices were built in 1 second.

Table 5.5 shows the number of concepts at the final step. The number of concepts in concept lattices varies from one to five times the number of concepts in AOC-posets, except for the HealthProtection lattice. In this case, there are 18 times more concepts in the lattice than in the AOC-poset. The concept number of Iceberg lattices is very low compared to the others, suggesting there are no large groups of objects with the same content.

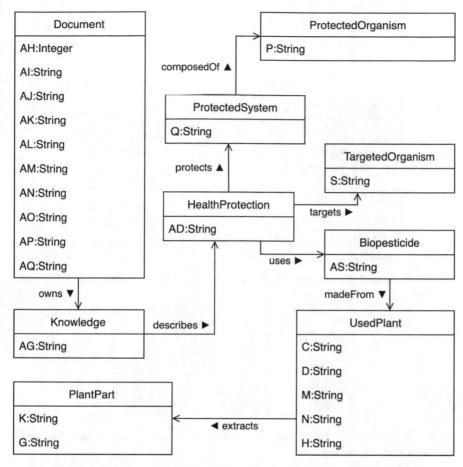

Fig. 5.6: KNOMANA: the UML model used for experimenting RCA algorithms

Table 5.6 shows the number of relational attributes at the final step. For a formal context *Source*, this number is related to the number of concepts of the ranges of the n_{out} relational contexts, i.e. *Range$_i$*, $1 \leq i \leq n_{out}$, leaving *Source*. For example, the Document relational attributes in the concept lattice (568) originated from the concepts of the Knowledge concept lattice. There was a cumulative effect for HealthProtection, the relational attributes in the concept lattice (9545) resulting from the union of concepts in ProtectedSystem (23), TargetedOrganism (936), and Biopesticide (8586) concept lattices. This number of relational attributes in HealthProtection concept lattice (9545) generated a high number of concepts in HealthProtection concept lattice (224992) due to a dispersion of descriptions. During the propagation along *describes* relation, this dispersion was absorbed: 568 concepts only in Knowledge concept lattice, and further re-expanded during propagation through *owns* relation in Document concept lattice (6272 concepts).

Formal context	#Objects	#original attributes	#Boolean attributes	density	Relational context	Density
Document	3541	10	6350	7.73 E-4	**Owns**	0.062
Knowledge	16	1	16	0.062	**Describes**	0.063
HealthProtection	10172	1	30	0.033	**Protects**	0.168
ProtectedSystem	6	1	6	0.167	**ComposedOf**	0.206
ProtectedOrganism	195	1	195	0.005		
					Targets	0.001
TargetedOrganism	934	1	934	0.001		
					Uses	3.2 E-4
Biopesticide	3127	1	30	0.033	**MadeFrom**	3.23 E-4
UsedPlant	4078	5	4353	7.37 E-4	**Extracts**	0.068
PlantPart	344	2	377	0.005		

Table 5.3: KNOMANA: Dimensions of the relational context family

	#steps			time (ms)	time (mn)
FCA	7		FCA	135582	[Sim] 2
CERES	7		CERES	45355	[Sim] 1
PLUTON	7		PLUTON	498299	[Sim] 8
HERMES	7		HERMES	362232	[Sim] 6
ICEBERG10	7		ICEBERG10	517	[Sim] 0.01
ICEBERG30	7		ICEBERG30	246	[Sim] 0.01
ICEBERG40	7		ICEBERG40	206	[Sim] 0.01

Table 5.4: KNOMANA: (left) Final step number and (right) computation time (milliseconds) and (minutes)

Formal Context	CONCEPT LATTICE	AOC-POSET	ICEBERG10	ICEBERG30	ICEBERG40
BioPesticide	8586	3503	18	11	11
HealthProtection	224992	12182	102	31	15
UsedPlant	5651	4708	8	5	5
ProtectedSystem	23	21	8	8	2
ProtectedOrganism	197	195	2	2	2
PlantPart	386	384	5	2	2
Knowledge	568	116	49	20	8
TargetedOrganism	936	934	2	2	2
Document	6272	4181	74	28	7
TOTAL	247611	26224	268	109	54

Table 5.5: KNOMANA: Number of concepts for each conceptual structure

Formal Context	CONCEPT LATTICE	AOC-POSET	ICEBERG10	ICEBERG30	ICEBERG40
BioPesticide	5651	4716	8	5	5
HealthProtection	9545	4458	28	21	15
UsedPlant	386	384	5	2	2
ProtectedSystem	197	195	2	2	2
ProtectedOrganism	0	0	0	0	0
PlantPart	0	0	0	0	0
Knowledge	224992	12464	102	31	15
TargetedOrganism	0	0	0	0	0
Document	568	116	49	20	8
TOTAL	241339	22333	194	81	47

Table 5.6: KNOMANA: Number of relational attributes for each conceptual structure

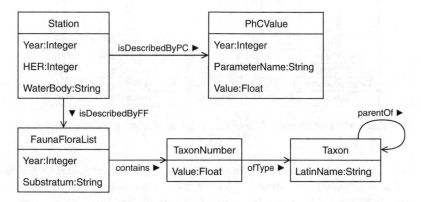

Fig. 5.7: FRESQUEAU: the UML model used for experimenting RCA algorithms

5.4.2.2 Experiments on FRESQUEAU Dataset

Figure 5.7 shows the UML model, with a loop, chosen for experimenting RCA algorithms on the FRESQUEAU database. As noted above, the dataset represents sampling sites (`Stations`) described by attributes and relations. A station is characterized by a `Year` of observation, it depends on a `WaterBody` and is located in a `HER` (i.e. a Hydro-Eco-Region). A station *is described by* annual average physico-chemical values (`PhCValue`) measured there for 22 parameters (`ParameterName`), and by some fauna or flora lists (`FaunaFloraList`) *containing* the numbers (`TaxonNumber`) of the various *types of* taxons (`Taxon`) collected there at most once a year. Furthermore, taxons have family relationships (`ParentOf`). The aim is to extract groups of stations having similar biological and physico-chemical characteristics.

Table 5.7 shows the dimensions of the RCF for the considered excerpt of FRESQUEAU database. This RCF has less contexts but they are larger than those of KNOMANA RCF. Here, the RCF is composed of 4 formal contexts and 5 relational contexts. The longest path (omitting the loop) in the UML model graph has 4 edges (from `Stations` to `Taxon`). The largest contexts are `TaxonNumber` (more than 120000 objects) and `PhCValue` (more than 18000 objects). Original attributes with float values have been discretized into quartiles. Nominal/integer original attributes generated many Boolean attributes, e.g. 383 for the `Station` context. Taxons are described with no attribute except their name, and thus, `Taxon` context is diagonal. Densities are most of the time low to very low, except for context `TaxonNumber` where there is only 4 Boolean attributes.

Formal context	#Objects	# original attributes	# Boolean attributes	# density	Relational context	# density
Station	4808	3	383	0.008	**isDescribedByPC**	0.004
PhCValue	18524	3	512	0.010	**isDescribedByFF**	0.001
FaunaFloraList	510	2	59	0.051	**contains**	1.41E-4
TaxonNumber	124242	1	4	0.25	**ofType**	2.94E-4
Taxon	3400	1	3389	2.95E-4	**parentOf**	2.15E-4

Table 5.7: FRESQUEAU: Dimensions of the relational context family

Table 5.8 shows the step number and the running time for the different algorithms applied to the FRESQUEAU dataset. Concept lattices construction with the add intent/extent algorithm (FCA) cannot finish due to a lack of memory. The RCA process converges in 5 steps for AOC-poset algorithms and Iceberg lattices construction with minimal support 10 (the 5th step being to confirm that the fixpoint is reached). Iceberg lattices construction with minimal supports 30 and 40 needs only 3 steps, due to the fact that less concepts are built and need to be propagated (see Table 5.9). AOC-poset construction takes between 4 and 25 minutes depending on the algorithms, CERES being again the most efficient. As for the KNOMANA dataset, Iceberg lattices are easily built in about 1 or 2 seconds.

	#steps		time (ms)	time (mn)
FCA	(-)	FCA	(-)	(-)
CERES	5	CERES	214577	[Sim] 4
PLUTON	5	PLUTON	1492604	[Sim] 25
HERMES	5	HERMES	979421	[Sim] 16
ICEBERG10	5	ICEBERG10	2152	[Sim] 0.04
ICEBERG30	3	ICEBERG30	892	[Sim] 0.01
ICEBERG40	3	ICEBERG40	711	[Sim] 0.01

Table 5.8: FRESQUEAU: (left) Final step number and (right) computation time (milliseconds) and (minutes)

Table 5.9 shows the number of concepts for each formal context at the final step. The AOC-poset with the highest number of concepts is the one built for `PhCValue` context. The concept number of Iceberg lattices is low or very low, suggesting, as for the KNOMANA dataset, that there are no large groups of objects with the same description.

Formal context	CONCEPT LATTICE	AOC-POSET	ICEBERG10	ICEBERG30	ICEBERG40
Station	(-)	1671	89	3	2
PhCValue	(-)	19013	7	2	2
FaunaFloraList	(-)	1171	110	5	5
TaxonNumber	(-)	4524	9	3	3
Taxon	(-)	3392	3	2	2
TOTAL	(-)	29771	218	15	14

Table 5.9: FRESQUEAU: number of concepts for each conceptual structure

Table 5.10 shows the number of relational attributes for each formal context at the final step. Context `PhCValue` has no relational attribute, not being the domain of any relation. `Station` context is extended with relational attributes pointing to either `PhCValue` or `FaunaFloraList` concepts. Extended contexts `Taxon` and `TaxonNumber` have the same number of relational attributes (the number of concepts of the corresponding `Taxon` lattice, see Table 5.9), due to the diagonality of `Taxon` context and the reflexivity of `parentOf` relation.

Formal Context	CONCEPT LATTICE	AOC-POSET	ICEBERG10	ICEBERG30	ICEBERG40
Station	(-)	20321	117	7	7
PhCValue	(-)	0	0	0	0
FaunaFloraList	(-)	4524	9	3	3
TaxonNumber	(-)	3392	3	2	2
Taxon	(-)	3392	3	2	2
TOTAL	(-)	31629	132	14	14

Table 5.10: FRESQUEAU: number of relational attributes for each conceptual structure

5.4.3 Discussion

In this section, we presented quantitative results about computation time and conceptual structure size for two datasets. We can learn lessons from these experiments on real environmental datasets. Their dimensions present differences, with a very huge object number in one of the formal contexts of FRESQUEAU (`TaxonNumber`). They also have some similarities, with low to very low densities in their formal and relational contexts. The presence of a loop in FRESQUEAU UML model may explain the impossibility to reach the fixpoint for concept lattice construction. Furthermore, when adding all the opposite relations in KNOMANA UML model, some conceptual structure computation reach the fixpoint: AOC-posets using CERES algorithm are computed within 25 steps in more than one hour and Iceberg 40 lattices are computed within 13 steps in less than a minute. Concept lattices can be built until step 3, Iceberg 10 lattices until step 4 and Iceberg 30 lattices until step 7. This suggests considering other computation strategies to assist domain experts during data exploration tasks: e.g. the process can be interrupted at a certain concept propagation step, which delivers knowledge patterns that can be sufficient for some investigations, or rather than building the whole conceptual structures, concepts can be built around a first focus concept, issued from a set of attributes or objects that have a particular interest for the experts [4].

Furthermore, considering that the conceptual structures are the informative search space for domain experts, raises the question of assisting experts in interpreting and drawing conclusions from the results. This can be made through summarizing patterns (as shown in next section), rule extraction [15, 42], or guided exploration of a focus concept neighbourhood [5]. Examples of domain questions for KNOMANA can be found in [29, 42]. Next section develops this question with the specific case of the FRESQUEAU project, giving insight on how it can be exploited by an expert to analyse relational data.

5.5 Analysing Sequences from Water Quality Monitoring Using RCA

In this section,we consider a smaller but complex relational dataset from FRESQUEAU database (see Sect. 5.4.1.2), that generates large number of concepts when processed by RCA. We show how the lattice family resulting from RCA can be summarized into a single lattice of graphs, to help the interpretation. The approach presented here can be generalized on any relational dataset, possibly larger, thanks to Iceberg lattices construction, and being provided a main lattice to start the summarizing process.

The approach was originally designed to help hydro-ecologists when analysing river water data, and trying to answer the following question: can sets of physico-chemical parameter values be temporally linked with bio-indicator values? To answer this question, Fabrègue et al. [18, 19] proposed to use a temporal pattern based method, extracting closed partially ordered patterns (CPO-patterns) from a sequence dataset. Following this idea, and to facilitate the analysis, RCA-Seq has been devised [39, 40] for extracting a hierarchy of CPO-patterns from the same datasets. The idea is to represent sequences within a relational context family, to build the lattice family, and then to transform a concept and its related concepts into a CPO-pattern, i.e. a directed acyclic graph (DAG), where each concept corresponds to a vertex, and each relational attribute to an edge. Thus, the lattice family is summarized into a hierarchy of concept graphs [22]. In the following we will explain the functioning of RCA-Seq and present some experiments. Finally we show how the obtained hierarchy can be used to help the expert analysis.

5.5.1 RCA-Seq

RCA-Seq spans three main steps: a) modelling sequential data, b) exploring sequential data with RCA and c) extracting DAGs (CPO-patterns) by navigating the RCA output.In the following, we concisely introduce them.

5.5.1.1 Modelling Hydro-Ecological Sequential Data.

In the following a sequence is a list of successive physico-chemical samples collected during a certain period before a biological sampling. To explore these sequential data with RCA, we use the model depicted in Fig. 5.8. The four rectangles represent the four sets of objects we manipulate, as follows: biological (Bio) samples, physico-chemical (PhC) samples, Bio indicators and PhC parameters. Let us note that the analysis only focuses on one Bio indicator at a time. The links between Bio/PhC samples and PhC samples are highlighted by the temporal binary relation *is preceded by*. This temporal relation associates one sample with another one if the first sample is preceded in time by the second one, on the same river stations. Data have been discretised, and the Bio/PhC samples are thus described only by the following binary quality relations *has parameter blue* (very good quality), *has parameter green* (good quality), *has parameter yellow* (medium quality), *has parameter orange* (bad quality) and *has parameter red* (very bad quality) that link the Bio/PhC samples with the measured Bio indicators/PhC parameters.

5.5.1.2 Exploring Hydro-Ecological Sequential Data with RCA

Firstly, based on the data model given in Fig. 5.8, all sequences, e.g. $Seq_1 = \langle \{NITR_{green}, PHOS_{green}\} \{NITR_{blue}\} \{NITR_{green}, PHOS_{blue}\} \{NITR_{green}, PHOS_{green}\} \{NITR_{green}\} \{IBGN_{green}\}\rangle$, are encoded into the RCA input as depicted in Table 5.11. The tables KPHC (PhC parameters), KBIOS (Bio samples) and KPHCS (PhC samples) represent object-attribute contexts. There is no object-attribute context for Bio indicators because all analysed sequences end with the same e.g. $IBGN_{green}$. KBIOS and KPHCS have no column since the samples are only described using the quality relations. Moreover they are identified based on the corresponding sequence. The tables RPHCS-ipb-PHCS, RBIOS-ipb-PHCS , RbPHC and RgPHC represent object-object contexts. For example, RPHCS-ipb-PHCS defines the temporal relations (ipb) between PhC samples and has KPHCS both as domain and range. RbPHC (resp. RgPHC) defines the quality relations between PhC samples and PhC parameters that have the blue (b) (resp. the green (g)) quality value.

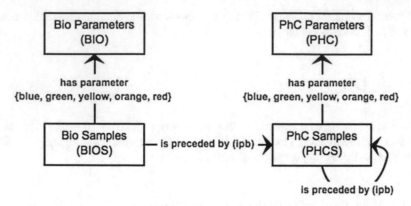

Fig. 5.8: The modelling [38] of hydro-ecological sequential data collected during the FRESQUEAU project; Bio and PhC stand respectively for biological and physico-chemical

object-attribute contexts

KPHC	NITR	PHOS
NITR	×	
PHOS		×

KBIOS	Seq$_1$
Seq$_1$	

KPHCS
J$_1$_Seq$_1$
J$_2$_Seq$_1$
J$_3$_Seq$_1$
J$_4$_Seq$_1$
J$_5$_Seq$_1$

object-object contexts

RBIOS-ipb-PHCS	J$_1$_Seq$_1$	J$_2$_Seq$_1$	J$_3$_Seq$_1$	J$_4$_Seq$_1$	J$_5$_Seq$_1$
Seq$_1$	×	×	×	×	×

RPHCS-ipb-PHCS	J$_1$_Seq$_1$	J$_2$_Seq$_1$	J$_3$_Seq$_1$	J$_4$_Seq$_1$	J$_5$_Seq$_1$
J$_1$_Seq$_1$					
J$_2$_Seq$_1$	×				
J$_3$_Seq$_1$	×	×			
J$_4$_Seq$_1$	×	×	×		
J$_5$_Seq$_1$	×	×	×	×	

RbPHC	NITR	PHOS
J$_1$_Seq$_1$		
J$_2$_Seq$_1$	×	
J$_3$_Seq$_1$		×
J$_4$_Seq$_1$	×	
J$_5$_Seq$_1$		×

RgPHC	NITR	PHOS
J$_1$_Seq$_1$	×	×
J$_2$_Seq$_1$		
J$_3$_Seq$_1$	×	
J$_4$_Seq$_1$	×	×
J$_5$_Seq$_1$	×	

Table 5.11: RCF composed of object-attribute contexts: KPHC, KBIOS and KPHCS; temporal object-object contexts: RBIOS-ipb-PHCS and RPHCS-ipb-PHCS; quality object-object contexts: RbPHC and RgPHC; J_k_Seq$_1$ uniquely identifies a PHC sample within a sequence Seq$_1$

Secondly, by applying RCA to an RCF as depicted in Table 5.11, three lattices are generated as follows: *the main lattice* $\mathscr{L}_{\text{KBIOS}} = (\mathscr{C}_{\text{KBIOS}}, \preceq_{\text{KBIOS}})$, *the temporal lattice* $\mathscr{L}_{\text{KPHCS}} = (\mathscr{C}_{\text{KPHCS}}, \preceq_{\text{KPHCS}})$ (i.e. PhC samples) and *the lattice of items* $\mathscr{L}_{\text{KPHC}} = (\mathscr{C}_{\text{KPHC}}, \preceq_{\text{KPHC}})$ (i.e. PhC parameters). Throughout the RCA process two types of relational attributes are defined based on the introduced binary relations, either on *time* or *quality*. When $C_m = (X_m, Y_m) \in \mathscr{C}_{\text{KBIOS}}$ has $\exists \text{RBIOS-ipb-PHCS}(C_{t1}) \in Y_m$, then C_m points to $C_{t1} \in \mathscr{C}_{\text{KPHCS}}$. When $C_{t1} = (X_{t1}, Y_{t1})$ has $\exists \text{RPHCS-ipb-PHCS}(C_{t2}) \in Y_{t1}$, then C_{t1} points to $C_{t2} \in \mathscr{C}_{\text{KPHCS}}$ (i.e. to another concept from the temporal lattice). In addition, when $\exists \text{RbPHC}(C_i) \in Y_{t1}$, then C_{t1} points to $C_i = (X_i, Y_i) \in \mathscr{C}_{\text{KPHC}}$. When $C_i \equiv \top(\mathscr{L}_{\text{KPHC}})$, X_i contains all PhC parameters; for other concepts X_i contains only one PhC parameter.

5.5.1.3 Extracting DAG by Navigating the RCA Output

As explained in [40], for each concept of the main lattice $C_m = (X_m, Y_m) \in \mathcal{L}_{\text{KBIOS}}$ is extracted a DAG (or CPO-pattern), namely $\mathcal{G}_{C_m}$, by navigating interrelated concept intents starting with Y_m; we define $Support(\mathcal{G}_{C_m}) = |X_m|$, i.e. the number of sequences that are summarized by $\mathcal{G}_{C_m}$. All generated DAG are directly organised into a hierarchy according to the inclusion on the associated main concept extents. This order corresponds to the subsumption on graphs $\prec_g$.

Briefly, we explain how to extract a DAG $\mathcal{G}_{C_m} = (\mathcal{V}_m, \mathcal{E}_m, l_m)$ (l_m is a labelling function) associated with a main concept $C_m = (X_m, Y_m) \in \mathcal{C}_{\text{KBIOS}}$ whose intent has at least one temporal relational attribute $\exists\text{RBIOS-ipb-PHCS}(C_{t1})$, where $C_{t1} = (X_{t1}, Y_{t1}) \in \mathcal{C}_{\text{KPHCS}}$. Concept C_m reveals a vertex $v_m \in \mathcal{V}_m$ labelled with an itemset containing the assessed Bio indicator, e.g $\{\text{IBGN}_{\text{green}}\}$. The aforementioned temporal relational attribute leads to another vertex $v_{t1} \in \mathcal{V}_m$ derived from C_{t1}, i.e. the edge $(v_{t1}, v_m) \in \mathcal{E}_m$ is disclosed. If a quality relational attribute $\exists\text{RbPHC}(C_i) \in Y_{t1}$ with $C_i = (X_i, Y_i) \in \mathcal{C}_{\text{KPHC}}$, then v_{t1} is labelled with Y_i. Precisely, if $C_i \equiv \top(\mathcal{L}_{\text{KPHC}})$, then the *abstract quality value* $?_{\text{blue}} \in l(v_{t1})$ is derived; if $C_i \prec_{\text{KPHC}} \top(\mathcal{L}_{\text{KPHC}})$ with e.g. $Y_i = \{\text{PHOS}\}$, then the *concrete quality value* $\text{PHOS}_{\text{blue}} \in l(v_{t1})$; if Y_{t1} has no quality relational attribute, then the *abstract value* $?_? \in l(v_{t1})$. If Y_{t1} contains a temporal relational attribute $\exists\text{RPHCS-ipb-PHCS}(C_{t2})$, then C_{t1} leads to another vertex $v_{t2} \in \mathcal{V}_m$ derived from C_{t2}. Therefore, the order on vertices in $\mathcal{G}_{C_m}$ is revealed by temporal relational attributes; the itemsets labelling the vertices are revealed by quality relational attributes. When all next navigated concept intents have no temporal relational attribute, then the extraction of $\mathcal{G}_{C_m}$ is finished.

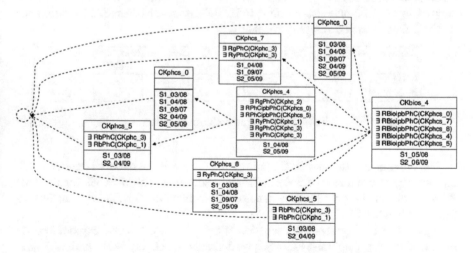

Fig. 5.9: Extracting a DAG by navigating the relational attributes starting from a KBIOS concept; the objects in the extents represent (station, time stamp) pairs, the intents of $C_{\text{KPHC\_N}}$ concepts are parameters, NITR (N=1), PHOS (2), any (3)

Figure 5.9 illustrates the extraction of a DAG starting from a concept $C_{\text{KBIOS\_4}}$ (right of the figure). This concept has 5 temporal relational attributes that lead to 5 concepts of the lattice $\mathscr{L}_{\text{KPHCS}}$. Some of these concepts have quality relational attributes (e.g. $\exists\,\text{RgPHC}(C_{\text{KPHC\_3}})$ for concept $C_{\text{KPHCS\_7}}$) leading to concepts of $\mathscr{L}_{\text{KPHC}}$; while others (e.g. $C_{\text{KPHCS\_4}}$) have temporal relational attributes that lead again to concepts of $\mathscr{L}_{\text{KPHCS}}$ (left of the figure). Since these last concepts ($C_{\text{KPHCS\_5}}$, $C_{\text{KPHCS\_0}}$) have no temporal relational attributes, the extraction is finished.

5.5.2 Experiments

This section presents an experimental study of our approach. The experiments were carried out on a MacBook Pro with a 2.9 GHz Intel Core i7, 8GB DDR3 RAM running OS X 10.9.5. The family lattice was built with RCAExplore.[4] The extraction step relied on the CPOHrchy algorithm from [40].

To assess the performance of RCA-Seq we used two hydro-ecological sequential datasets, IBD blue and IBGN blue, whose characteristics, i.e, number of sequences, number of PHC samples, number of PHC parameters, average sequence length (the number of PHC samples in the sequence), maximum sequence length and density, are shown in Table 5.12. Figure 5.10a and d depict the number of obtained DAGs (vertical axis) with respect to the minimum support θ (%) (horizontal axis) in the IBD and IBGN blue dataset. Even if both datasets have almost the same number of sequences, the extracted number of DAGs varies. For instance, 300411 DAGs are discovered in the IBGN blue dataset with $\theta = 9\%$, while for the same minimum support in the IBD blue dataset only 16525 DAGs are discovered. This difference can be linked to each dataset heterogeneity.

Dataset	#sequences	#PHC samples	#PHC parameters	Avg. seq. length	Max. seq. length	Density
IBD blue	1196	3012	46	2.51	7	2.37E−4
IBGN blue	1102	3077	26	2.79	8	3.17E−4

Table 5.12: Dataset characteristics

The number of extracted DAGs is important, even if the dataset is rather small, e.g. we report a number of 569202 DAGs discovered with $\theta = 3\%$ for the IBD dataset that contains only 1196 sequences built from 46 items and having an average sequence length of 2.51 (Fig. 5.10a).

Figure 5.10b illustrates the execution time of the RCA-based exploration. As explained in [40], to optimise RCA-Seq we defined respectively for the lattices $\mathscr{L}_{\text{KBIOS}}$ and $\mathscr{L}_{\text{KPHCS}}$ the minimum supports θ and θ', where $\theta' = \theta\frac{|G_{\text{KBIOS}}|}{|G_{\text{KPHCS}}|}$. G_{KBIOS} and G_{KPHCS} are respectively the set of objects of the KBIOS and KPHCS object-attribute contexts. For instance, when θ' is not defined (i.e. non-optimised RCA-exploration), during

---

[4] http://dataqual.engees.unistra.fr/logiciels/rcaExplore.

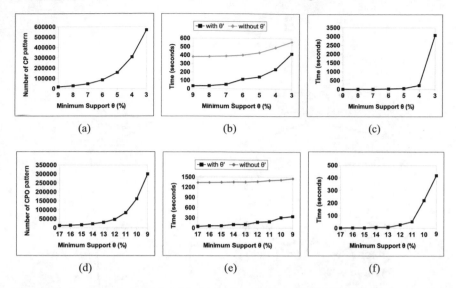

Fig. 5.10: Performance evaluation based on IBD and IBGN blue datasets of Table 5.12; minimum support θ is defined for $\mathscr{L}_{\text{KBIOS}}$; θ' is defined for $\mathscr{L}_{\text{KPHCS}}$. (**a**) DAGs (IBD). (**b**) RCA-exploration. (**c**) CPOHrchy. (**d**) DAGs (IBGN). (**e**) RCA-exploration. (**f**) CPOHrchy

the iterative steps the relational scaling mechanism processes $|\mathscr{C}_{\text{KPHCS}}| = 105850$ temporal concepts even if not all of them are used to extract DAGs. When $\theta = 6\%$ and $\theta' = 3\%$, only $|\mathscr{C}_{\text{KPHCS}}| = 4429$ temporal concepts are generated; when $\theta = 3\%$ and $\theta' = 1\%$, then $|\mathscr{C}_{\text{KPHCS}}| = 31854$ temporal concepts are generated. Thus, for $\theta = 6\%$ and $\theta = 3\%$ the optimised RCA-based exploration is respectively 3.49 and 1.33 times faster than the non-optimised one.

Figure 5.10c shows the computation time of the algorithm CPOHrchy. It is noted that low values of θ, i.e. $< 4\%$, and high numbers of DAGs, i.e. ≥ 300000, slow down the extraction step. In addition, the efficiency of CPOHrchy can be influenced by the used implementation,[5] which is not currently optimised for searching in large collections.

5.5.3 Navigating the Resulting Hierarchy of Graphs

Figure 5.11 depicts an excerpt from a hierarchy of DAGs extracted by applying RCA-Seq to an IBGN blue dataset with 80 analysed hydro-ecological sequences. This excerpt highlights two benefits of exploring qualitative sequential data by means of RCA. Firstly, the generalisation order regarding the structure of the extracted DAGs.

---

[5] Based on Java Collection Framework and Lambda Expressions.

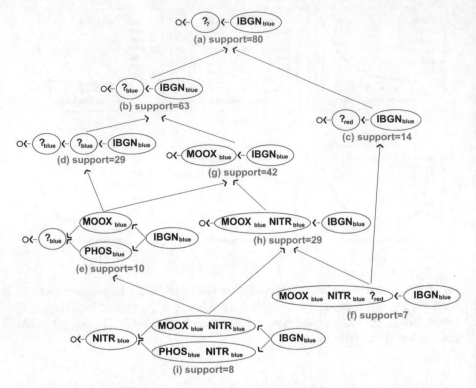

Fig. 5.11: Excerpt from a hierarchy of DAGs generated from an IBGN blue dataset with 80 analysed hydro-ecological sequences

For example, the structure of DAG (e) is more specific than the structure of its ancestor DAGs, i.e. there exists a projection from its ancestor DAGs into (e). Secondly, the partial order on items, and the inclusion order on itemsets. For instance, DAG (g) reveals the regularity $\{MOOX_{blue}\} \leftarrow \{IBGN_{blue}\}$ (i.e. a blue quality of IBGN Bio indicator is frequently preceded by a blue quality of MOOX PhC parameter) that is a specialisation of the less accurate regularity $\{?_{blue}\} \leftarrow \{IBGN_{blue}\}$ (i.e. a blue quality of IBGN Bio indicator is frequently preceded by a blue quality of some PhC parameter) revealed by DAG (b). Similarly, DAG (i) reveals $\{PHOS_{blue}, NITR_{blue}\} \leftarrow \{IBGN_{blue}\}$ regularity that is a specialisation of the regularity $\{PHOS_{blue}\} \leftarrow \{IBGN_{blue}\}$ revealed by (e). In addition, DAG (e), having 12.5% frequency (i.e. in Fig. 5.11, DAG (e) has *Support* = 10, and the total number of analysed sequences is equal to 80), can be found when, e.g. $\theta = 12\%$, even if its accurate specialization DAG (i) is not frequent, and thus is not extracted. These properties of the extracted hierarchies help experts in understanding the obtained knowledge, and, besides, provide a quick way to navigate to interesting DAGs.

Accordingly, the hierarchy in Fig. 5.11 can be navigated starting from the more general DAGs. Thus, the experts have an overview of the trends within analysed data, and minimize the chance of overlooking interesting ones. DAG (a) confirms that all analysed Bio samples are preceded by at least one PhC sample. Both direct descendants, the (b) and (c) DAGs, emphasize two well-known correspondence between the qualities of PhC parameters and the ones of IBGN Bio indicator. Firstly, DAG (b), which is retrieved with 78.75% frequency in the analysed data, highlights that $IBGN_{blue}$ is frequently preceded by PhC parameters having *blue* qualities. Secondly, DAG (c), which is retrieved in 17.5% of the analysed data, stresses that *red* PhC parameters are not frequently measured before an $IBGN_{blue}$ since they produce a degradation of the watercourse qualities, and do not lead to a very good ecological status. As expected, in contrast with DAG (b), DAG (c) has a low support. Therefore, the experts can navigate only descendants of DAG (b) in order to find patterns revealing pertinent synergies between PhC parameters and $IBGN_{blue}$ that provide a very good quality of watercourses. In addition, the experts can focus only on descendants of (c) to find out how the watercourse degradation is neutralised when *red* PhC parameters are measured. For example, the neutralisation of $?_{red}$ is possible when these very bad values of PhC parameters coexist at the same time with the good values $MOOX_{blue}$ and $NITR_{blue}$ as shown by DAG (f). Following the same principles, the experts can continue the navigation being guided by the relationships between the extracted DAGs and the information about their support.

5.6 Conclusion

In this paper, we have presented results on the application of RCA to environmental datasets coming from the real world and built under guidance of domain experts. The two application domains are biopesticides and antimicrobial products made from plants and assessment of the quality of waterbodies. We have shown the scope of the RCA process in terms of quantitative opportunities and limits on our datasets. We also have described qualitative results in the second domain, about the challenging issue of temporally linking physico-chemical parameter values with bio-indicator values.

We are pursuing two related main tracks of research: (1) improving time and space efficiency of the RCA implementation and adding new algorithmic strategies; (2) improving guidance of experts in their analysis.

Concerning track (1), finding opportunities for *space and time efficiency improvement* is a main and complex task, with tangled concerns, both theoretical and technical. We are experimenting various collection types as many libraries exist that can have an impact on efficiency: Java API collections (currently `BitSet` is used as a main provider for efficiency in experiments of Sect. 5.4), colt library[6] (used in experiments of Sect. 5.5), Apache common collection library,[7] Google Guava,[8] etc. Other

---

[6] https://dst.lbl.gov/ACSSoftware/colt/.

[7] https://commons.apache.org/proper/commons-collections/.

[8] https://github.com/google/guava/wiki.

construction algorithms for the concept lattice will be implemented as well, such as described and experimented in [1, 54]. Another technical but important concern is about the data, whose input and output file format (currently textual input format, and dot output format, with optionally XML output file unfeasible on large results) and memory encoding (currently adjacency lists) have an impact on efficiency. We also are designing on-demand and local algorithms for RCA, following the first work presented in [5]. There are plenty of different ways to consider an on-demand local algorithm, and this way of computing and delivering results has a strong potential for complex, large and evolving datasets.

Concerning track (2), *guidance of experts* can be strenghten by various means. RCA quantifiers offer many possibilities for analysis, with the counterpart that the expert may be lost when choosing parameters (quantifiers and conceptual structures). To that aim, we are studying assisting methodologies, by providing a controlled language for expressing a general "query" with different quantifiers associated with the various relations, and the possible choice of only parts of the RCF; by controlling the coherence between the quantifier choices on semantically connected relations; and by anticipating the result size on neighbouring configurations (with slight changes in the analysed RCF part, or with "similar" quantifiers). The other challenge is providing a user interface with result visualization adapted to the domain experts. Presenting concept orders is used in many tools, while others focus on a particular concept and allow navigating to its neighbours [20], or give an alternative view on the conceptual structure through tag clouds [27].

Acknowledgements This work was supported by the French National Research Agency: (1) FRESQUEAU project referred as ANR11_MONU14; (2) KNOMANA project under the Investments for the Future Program, #Digitag, referred as ANR-16-CONV-0004. KNOMANA project is also supported by INRA-CIRAD GloFoodS metaprogram. FRESQUEAU database was completed thanks to the support of French Office of Biodiversity (OFB). We warmly acknowledge Corinne Grac (UMR7362 LIVE - ENGEES) for her advice about the FRESQUEAU datasets.

References

1. Andrews, S.: Making use of empty intersections to improve the performance of cbo-type algorithms. In: Formal Concept Analysis - 14th International Conference, ICFCA 2017, Rennes, France, June 13–16, 2017, Proceedings, pp. 56–71 (2017). https://doi.org/10.1007/978-3-319-59271-8_4
2. Association Française de Normalisation: Qualité de l'eau : détermination de l'Indice Biologique Diatomées (IBD). NF T90-354 (2003)
3. Association Française de Normalisation: Qualité de l'eau : détermination de l'Indice Biologique Global Normalisé (IBGN). XP T90-350 (2004)
4. Bazin, A., Carbonnel, J., Huchard, M., Kahn, G.: On-demand relational concept analysis. CoRR **abs/1803.07847** (2018). URL http://arxiv.org/abs/1803.07847
5. Bazin, A., Carbonnel, J., Huchard, M., Kahn, G., Keip, P., Ouzerdine, A.: On-demand relational concept analysis. In: Formal Concept Analysis - 15th

International Conference, ICFCA 2019, Frankfurt, Germany, June 25–28, 2019, Proceedings, pp. 155–172 (2019). https://doi.org/10.1007/978-3-030-21462-3_11

6. Belohlavek, R., Macko, J.: Selecting important concepts using weights. In: P. Valtchev, R. Jäschke (eds.) Formal Concept Analysis: 9th International Conference, ICFCA 2011, Nicosia, Cyprus, May 2–6, 2011. Proceedings, pp. 65–80. Springer Berlin Heidelberg (2011)

7. Berry, A., Gutierrez, A., Huchard, M., Napoli, A., Sigayret, A.: Hermes: a simple and efficient algorithm for building the AOC-poset of a binary relation. Ann. Math. Artif. Intell. **72**(1–2), 45–71 (2014). https://doi.org/10.1007/s10472-014-9418-6

8. Berry, A., Huchard, M., McConnell, R.M., Sigayret, A., Spinrad, J.P.: Efficiently computing a linear extension of the sub-hierarchy of a concept lattice. In: B. Ganter, R. Godin (eds.) Formal Concept Analysis, Third International Conference, ICFCA 2005, Lens, France, February 14–18, 2005, Proceedings, *Lecture Notes in Computer Science*, vol. 3403, pp. 208–222. Springer (2005). https://doi.org/10.1007/978-3-540-32262-7_14

9. Braud, A., Dolques, X., Huchard, M., Le Ber, F.: Generalization effect of quantifiers in a classification based on relational concept analysis. Knowledge-Based Systems **160**, 119–135 (2018)

10. Buzmakov, A., Egho, E., Jay, N., Kuznetsov, S.O., Napoli, A., Raïssi, C.: On Mining Complex Sequential Data by means of FCA and Pattern Structures. Int. Journal of General Systems **45**, 135–159 (2016)

11. Buzmakov, A., Kuznetsov, S.O., Napoli, A.: Is Concept Stability a Measure for Pattern Selection? Procedia Computer Science **31**, 918–927 (2014)

12. Codocedo, V., Bosc, G., Kaytoue, M., Boulicaut, J.F., Napoli, A.: A proposition for sequence mining using pattern structures. In: Proceedings of the 14th Int. Conf. on Formal Concept Analysis, ICFCA, pp. 106–121. Springer (2017)

13. De Maio, C., Fenza, G., Gallo, M., Loia, V., Senatore, S.: Formal and relational concept analysis for fuzzy-based automatic semantic annotation. Applied Intelligence **40**(1), 154–177 (2014)

14. Dolques, X., Huchard, M., Nebut, C., Reitz, P.: Fixing Generalization Defects in UML Use Case Diagrams. Fundam. Inform. **115**(4), 327–356 (2012)

15. Dolques, X., Le Ber, F., Huchard, M., Grac, C.: Performance-friendly rule extraction in large water data-sets with AOC posets and relational concept analysis. Int. J. General Systems **45**(2), 187–210 (2016). https://doi.org/10.1080/03081079.2015.1072927

16. Ducrou, J., Eklund, P.: SearchSleuth: The Conceptual Neighbourhood of an Web Query. In: CEUR Workshop Proceedings: Concept Lattices and their Applications, vol. 331, pp. 249–259 (2007)

17. Džeroski, S.: Multi-relational data mining: an introduction. ACM SIGKDD Explorations Newsletter **5**(1), 1–16 (2003)

18. Fabrègue, M., Braud, A., Bringay, S., Grac, C., Le Ber, F., Levet, D., Teisseire, M.: Discriminant temporal patterns for linking physico-chemistry and biology in hydro-ecosystem assessment. Ecological Informatics **24**, 210–

221 (2014). https://doi.org/10.1016/j.ecoinf.2014.09.003. URL https://hal.archives-ouvertes.fr/hal-01090331

19. Fabrègue, M., Braud, A., Bringay, S., Le Ber, F., Teisseire, M.: Mining closed partially ordered patterns, a new optimized algorithm. Knowledge-Based Systems **79**, 68 – 79 (2015). https://doi.org/10.1016/j.knosys.2014.12.027. URL http://www.sciencedirect.com/science/article/pii/S0950705114004730

20. Ferré, S.: Camelis: a logical information system to organise and browse a collection of documents. Int. J. General Systems **38**(4), 379–403 (2009). https://doi.org/10.1080/03081070902857886

21. Ferré, S.: A Proposal for Extending Formal Concept Analysis to Knowledge Graphs. In: 13th Int. Conference, ICFCA 2015, Nerja, Spain, LNCS 9113, pp. 271–286 (2015)

22. Ferré, S., Cellier, P.: How hierarchies of concept graphs can facilitate the interpretation of RCA lattices? In: 14th Int. Conference CLA 2018, Olomouc, Czech Republic, pp. 69–80 (2018)

23. Ferré, S., Ridoux, O., Sigonneau, B.: Arbitrary Relations in Formal Concept Analysis and Logical Information Systems. In: 13th Int. Conf. on Conceptual Structures, ICCS'05, Kassel, Germany, LNAI 3596, pp. 166–180. Springer (2005)

24. Ganter, B., Kuznetsov, S.O.: Pattern structures and their projections. In: 9th Int. Conference ICCS'01, Stanford, CA, USA, pp. 129–142 (2001)

25. Ganter, B., Wille, R.: Formal concept analysis - mathematical foundations. Springer (1999)

26. Gizdatullin, D., Ignatov, D.I., Mitrafanova, E., Muratova, A.: Classification of demographic sequence basd on pattern structures and emerging patterns. In: Supplementary Proceedings of ICFCA, pp. 49–66 (2017)

27. Greene, G.J., Esterhuizen, M., Fischer, B.: Visualizing and exploring software version control repositories using interactive tag clouds over formal concept lattices. Information & Software Technology **87**, 223–241 (2017). https://doi.org/10.1016/j.infsof.2016.12.001

28. Hacene, M.R., Huchard, M., Napoli, A., Valtchev, P.: Relational concept analysis: mining concept lattices from multi-relational data. Ann. Math. Artif. Intell. **67**(1), 81–108 (2013)

29. Keip, P., Gutierrez, A., Huchard, M., Le Ber, F., Sarter, S., Silvie, P., Martin, P.: Effects of input data formalisation in relational concept analysis for a data model with a ternary relation. In: Formal Concept Analysis - 15th International Conference, ICFCA 2019, Frankfurt, Germany, June 25–28, 2019, Proceedings, pp. 191–207 (2019). https://doi.org/10.1007/978-3-030-21462-3_13

30. Kötters, J.: Concept Lattices of a Relational Structure. In: 20th Int. Conf. ICCS 2013, Mumbai, India, LNCS 7735, pp. 301–310 (2013)

31. Krmelova, M., Trnecka, M.: Boolean Factor Analysis of Multi-Relational Data. In: CLA 2013, La Rochelle, France, CEUR Workshop Proc. 1062, pp. 187–198 (2013)

32. Leblanc, H.: Sous-hiérarchie de Galois : un modèle pour la construction et l'évolution des hiérarchies d'objets. Ph.D. thesis, Université de Montpellier (2000)
33. Lehmann, F., Wille, R.: A triadic approach to formal concept analysis. In: 3rd Int. Conference ICCS'95, Santa Cruz, California, USA, pp. 32–43 (1995)
34. Liquière, M., Sallantin, J.: Structural Machine Learning with Galois Lattice and Graphs. In: ICML, Madison, Wisconsin, pp. 305–313 (1998)
35. Martin, P., Sarter, S., Tagne, A., Ilboudo, Z., Marnotte, P., Silvie, P.: Knowing the useful plants for organic agriculture according to literature: Building and exploring a knowledge base for plant and animal health. In: African organic conference, pp. 137–141 (2018)
36. van der Merwe, D., Obiedkov, S.A., Kourie, D.G.: Addintent: A new incremental algorithm for constructing concept lattices. In: 2nd Int. Conference ICFCA 2004, Sydney, Australia, pp. 372–385 (2004)
37. Miralles, A., Molla, G., Huchard, M., Nebut, C., Deruelle, L., Derras, M.: Class model normalization - outperforming formal concept analysis approaches with aoc-posets. In: Proceedings of the Twelfth International Conference on Concept Lattices and Their Applications, Clermont-Ferrand, France, October 13–16, 2015, pp. 111–122 (2015). URL http://ceur-ws.org/Vol-1466/paper09.pdf
38. Nica, C., Braud, A., Dolques, X., Huchard, M., Le Ber, F.: Exploring Temporal Data Using Relational Concept Analysis: An Application to Hydroecology. In: M. Huchard, S. Kuznetsov (eds.) CLA: Concept Lattices and their Applications, *CEUR Workshop Proceedings*, vol. 1624, pp. 299–311. Moscow, Russia (2016). URL https://hal.archives-ouvertes.fr/hal-01380404
39. Nica, C., Braud, A., Dolques, X., Huchard, M., Le Ber, F.: Extracting Hierarchies of Closed Partially-Ordered Patterns Using Relational Concept Analysis. In: 22nd Int. Conf. ICCS 2016, Annecy, France, pp. 17–30 (2016)
40. Nica, C., Braud, A., Le Ber, F.: RCA-Seq: an Original Approach for Enhancing the Analysis of Sequential Data Based on Hierarchies of Multilevel Closed Partially-Ordered Patterns. Discrete Applied Mathematics **273**, 232–251 (2020). https://doi.org/10.1016/j.dam.2019.02.037
41. O'Neill, J.: Tackling drug-resistant infections globally: final report and recommendations. Review on Antimicrobial Resistance. Wellcome Trust and the Department of Health of United Kingdom (2016). 80 pages
42. Ouzerdine, A., Braud, A., Dolques, X., Huchard, M., Le Ber, F.: Adjusting the exploration flow in Relational Concept Analysis. An experience on a watercourse quality dataset. In: R. Jaziri, A. Martin, M.- Christine Rousset, L. Boudjeloud Assala, F. Guillet (eds.) Advances in Knowledge Discovery and Management: Volume 9, Studies in Computational Intelligence, vol. 1004, Springer Nature Switzerland AG, March 2022. ISBN 978-3030902865
43. Poelmans, J., Kuznetsov, S.O., Ignatov, D.I., Dedene, G.: Formal Concept Analysis in knowledge processing: A survey on models and techniques. Expert Syst. Appl. **40**(16), 6601–6623 (2013). https://doi.org/10.1016/j.eswa.2013.05.007

44. Prediger, S., Wille, R.: The Lattice of Concept Graphs of a Relationally Scaled Context. In: 7th Int. Conf. on Conceptual Structures, ICCS'99, Blacksburg, Virginia, LNCS 1640, pp. 401–414. Springer (1999)
45. Priss, U.: Relational concept analysis: Semantic structures in dictionaries and lexical databases. Ph.D. thesis, Technische Universität Darmstadt (1996)
46. Priss, U.: Formal concept analysis in information science. ARIST **40**(1), 521–543 (2006). https://doi.org/10.1002/aris.1440400120
47. Shi, L., Toussaint, Y., Napoli, A., Blansché, A.: Mining for Reengineering: An Application to Semantic Wikis Using Formal and Relational Concept Analysis, pp. 421–435. Springer Berlin Heidelberg, Berlin, Heidelberg (2011)
48. Singh, P.K.: m-polar fuzzy graph representation of concept lattice. Engineering Applications of Artificial Intelligence **67**(Supplement C), 52–62 (2018)
49. Stumme, G.: Efficient data mining based on formal concept analysis. In: Database and Expert Systems Applications, pp. 534–546. Springer (2002)
50. Stumme, G., Taouil, R., Bastide, Y., Pasquier, N., Lakhal, L.: Computing iceberg concept lattices with Titanic. Data & Knowledge Engineering **42**(2), 189–222 (2002)
51. The European Parliament and the Council: Framework for Community action in the field of water policy. Directive 2000/60/EC (2000)
52. Voutsadakis, G.: Polyadic concept analysis. Order **19**(3), 295–304 (2002). https://doi.org/10.1023/A:1021252203599
53. Wolff, K.E.: Relational scaling in relational semantic systems. In: Conceptual Structures: Leveraging Semantic Technologies, 17th International Conference on Conceptual Structures, ICCS 2009, Moscow, Russia, July 26–31, 2009. Proceedings, pp. 307–320 (2009). https://doi.org/10.1007/978-3-642-03079-6_24
54. Wray, T., Outrata, J., Eklund, P.W.: Scalable Performance of FCbO Algorithm on Museum Data. In: Proceedings of the Thirteenth International Conference on Concept Lattices and Their Applications, Moscow, Russia, July 18–22, 2016, pp. 363–376 (2016). URL http://ceur-ws.org/Vol-1624/paper28.pdf
55. Wu, W.Z., Leung, Y., Mi, J.S.: Granular computing and knowledge reduction in formal contexts. IEEE Transactions on Knowledge and Data Engineering **21**(10), 1461–1474 (2009)

Chapter 6
Computing Dependencies Using FCA

Jaume Baixeries, Victor Codocedo, Mehdi Kaytoue, and Amedeo Napoli

6.1 Introduction

The goal of this paper is to present and discuss some of our recent advances on the characterization and computation of different database constraints with Formal Concept Analysis [24] enriched with the formalism of pattern structures [23].

Although the Relational Database Model (RDBM) [31] currently coexists with other non-SQL database models [22], it still is one of the most popular database systems nowadays. In this model, one of the most important constraints is based on **Functional Dependencies** [30]. These dependencies are of key importance in the normalization of a database schema (the process of splitting an original dataset into smaller units in order to prevent redundancies and anomalies in the update process), and also in some accessory tasks, like data cleaning [15].

However, the definition of FDs is too strict for several useful tasks that are not related to the design of a database. One prototypical example is when one should model datasets with imprecision. By this, we mean datasets that contain errors and uncertainty in real-world data. To overcome this problem, different generalizations of FDs have been defined. These generalizations can be classified according to the criteria by which they relax the equality condition of FDs [16]. According to this

J. Baixeries (✉)
Computer Science Department, Universitat Politècnica de Catalunya, Barcelona, Spain
e-mail: jbaixer@cs.upc.edu

V. Codocedo
Departamento de Informática, Universidad Técnica Federico Santa María, Campus San Joaquín, Santiago de Chile, Chile
e-mail: victor.codocedo@inf.utfsm.cl

M. Kaytoue
Infologic R&D, Bourg-Lès-Valence, France
e-mail: mehdi.kaytoue@insa-lyon.fr

A. Napoli
Université de Lorraine, CNRS, Inria, LORIA, Nancy, France
e-mail: amedeo.napoli@loria.fr

© The Author(s), under exclusive license to Springer Nature Switzerland AG 2022
R. Missaoui et al. (eds.), *Complex Data Analytics with Formal Concept Analysis*,
https://doi.org/10.1007/978-3-030-93278-7_6

classification, two main strategies are presented: "extent relaxation" and "attribute relaxation" (in agreement with the terminology introduced in [16]).

The difference between these two strategies is that in the case of *extent relaxation*, a functional dependency may hold *only* in a (predefined) fraction of the datasets. In the *attribute relaxation* case, the equality that holds in the definition of a functional dependency may be *relaxed*, this is, it can be generalized by a more relaxed and general definition of an equality. The dependencies that follow this latter strategy have been named in different ways: fuzzy FDs [12], matching dependencies [33], constraint generating dependencies [13]. We name them **Similarity Dependencies**. These dependencies are still an active topic of research in the database community [14, 21, 33, 34].

Formal Concept Analysis (FCA) is a branch of applied lattice theory that dates back to the 1980s [24]. It is based on a binary relation between a set of objects and a set of attributes, and constructs a set of formal concepts with two operators between objects and attributes that form a Galois connection. Formal Concept Analysis has proved to be useful in many different fields: artificial intelligence, knowledge management, data-mining and machine learning, morphological mathematics, etc. [27].

Pattern structures [28] are a generalization of FCA for dealing with complex data within the framework of FCA [26, 28]. One of the interesting features of pattern structures is that they do not need to transform the original dataset before any treatment. As we will later see, this is one keypoint of the results presented here.

The characterization of FDs is a fruitful field in lattice theory [17, 20, 29]. Since the characterization of FDs with Formal Concept Analysis (FCA) was originally presented in [24], many papers have continued this line of research. Although this list is not intended to be exhaustive, we name the following general lines of research:

- Characterization of Functional Dependencies [2, 29].
- Characterization of Similarity Dependencies [12, 29].
- Characterization of other Database Dependencies [4–6, 9, 32].

In this paper, we present some recent results that deal with the characterization of FDs and similarity dependencies in terms of FCA and Pattern Structures, which have already been published [8–11]. The aim of this paper is not only to offer a general overview of the results obtained by these authors, but to try to present a more general overview of this topic, as well as a discussion of the limitations and potential possibilities that this line of research offers to both researchers and practitioners.

This line of research is relevant to different fields of research, namely, the database dependencies computation in the relational model, since we provide the possibility that existing algorithms in FCA may be used in the relational database model, or complex database analysis, since the generalization of functional dependencies allows the modeling of more complex relations within a dataset.

We present first the basic notations in Sect. 6.2, and then, we present some previous results that link FCA with the characterization of FDs. We introduce the results in Sect. 6.4, which are later discussed in Sect. 6.5.

6.2 Notation

In this section, we present the notation that will be used in this paper. The main object that will be used is a **dataset** D (equivalently **table, dataset, set of tuples**), that is composed by a set of **attributes** $\mathcal{U}$ and a set of **tuples** T, this is: $D = (\mathcal{U}, T)$. Along the paper, we may refer to either a dataset $D = (\mathcal{U}, T)$ or to the set of tuples T that it contains, in which case, we assume also the set of attributes $\mathcal{U}$.

The domain of $\mathcal{U}$ is *Dom*, which is a set of values. In the forthcoming examples, we assume that *Dom* is a numerical set. A tuple t is a function $t : \mathcal{U} \mapsto Dom$. Usually tables are presented as in Table 6.1, where the set of tuples (or objects) is $T = \{t_1, t_2, t_3, t_4\}$, $\mathcal{U} = \{a, b, c, d\}$ is the set of attributes, and $Dom = \{1, 3, 4, 7, 8\}$. Given a tuple $t \in T$, we say that $t(X)$ (for all $X \subseteq \mathcal{U}$) is a tuple with the values of t in the attributes $x_i \in X$:

$$t(X) = \langle t(x_1), t(x_2), \ldots, t(x_n) \rangle$$

For example, we have that $t_2(\{a, c\}) = \langle t_2(a), t_2(c) \rangle = \langle 4, 4 \rangle$. We also drop the set notation: instead of $\{a, b\}$ we use ab.

id	a	b	c	d
t_1	1	3	4	1
t_2	4	3	4	3
t_3	1	8	4	1
t_4	4	3	7	3

Table 6.1: An example of a table D, i.e. a set of tuples

6.2.1 Equivalence Relation

An equivalence relation R on a set T can be seen as:

1. A set of pairs, since $R \subseteq T \times T$.
2. A partition of the set T, this is, a set of subsets of $P = \{S_1, S_2, \ldots, S_n\}$, such that each pair $S_i \cap S_j = \emptyset$ when $i \neq j$ and $T = S_1 \cup S_2 \cup \cdots \cup S_n$. Each S_i is a **class** of T.

For example, given $P = \{\{1, 2, 3\}, \{4\}\}$, one has the relation

$$R_P = \{(1,2), (1,3), (2,3), (1,1), (2,2), (3,3), (4,4)\}$$

(omitting symmetry for the sake of readability). Equivalence relations contain also a partial order with the set inclusion operator when the equivalence relation is seen as a set of pairs, where the **meet** and **join** of two equivalence relations are the intersection and the union respectively. If this relation is seen as partitions of a set, the order if defined by the **finer** (**coarser**) relation: a partition P_i is finer than a partition P_j if every class of P_i is a subset of a class of P_j. The coarser relation is just the reverse.

An equivalence relation also defines a relation of two tuples w.r.t. a set of attributes. Following the notation in Example 6.4, we say that two tuples t_i, t_j are related w.r.t. the set of attributes X (this is: $t_i \theta_X t_j$) if $t_i(X) = t_j(X)$.

Example 6.1 We take tuples t_1, t_2 in Table 6.1, we state that $t_1 \theta_{bc} t_2$, but we do not have that $t_1 \theta_{ad} t_2$.

In this case, a set of attributes X generates an equivalence class Π_X.

Example 6.2 Again in Table 6.1, we state that $\Pi_{bc} = \{\{t_1, t_2\}, \{t_3\}, \{t_4\}\}$, and that $\Pi_a = \{\{t_1, t_3\}, \{t_2, t_4\}\}$.

6.2.2 Tolerance Relations

We introduce the concept of tolerance relation between values, which is a generalization of an equivalence relation. These two relations are extended to tuples and then used to characterize similarity and functional dependencies, respectively.

Definition 6.1 A *tolerance relation* $\theta \subseteq T \times T$ on a set T is a **reflexive** (i.e. $\forall t \in T : t \theta t$) and **symmetric** (i.e. $\forall t_i, t_j \in T : t_i \theta t_j \iff t_j \theta t_i$) relation.

A tolerance relation is somehow equivalent to the concept of **similarity** between two values, and this is how it will be used in this article.

Example 6.3 An often used tolerance relation is the *similarity* that can be defined within a set of integer values. Given two integer values v_1, v_2 and a user-defined threshold ε: $v_1 \theta v_2 \iff |v_1 - v_2| \leq \varepsilon$.

For instance, let's take a variable *Month*, defined over the months of a year. The function $\Delta_{Month}(m_1, m_2)$ defines the tolerance relation θ_{Month} such as:

$$\Delta_{Month}(m_1, m_2) = min(|m_1 - m_2|, min(m_1, m_2) + 12 - max(m_1, m_2))$$

$$m_i \theta_{Month} m_j \iff \Delta_{Month}(m_i, m_j) \leq 4$$

Then θ_{Month} is the tolerance relation that considers two values similar if they have values within 4 months of distance.

This tolerance relation on values can easily be extended to tuples.

Example 6.4 We now assume that the variable *Month* is, in fact, an attribute of a table dataset $D = (\mathscr{U}, T)$. The function $\Delta_{Month}(t_i, t_j)$, where $t_i, t_j \in T$, defines the tolerance relation θ_{Month} such as:

$$\Delta_{Month}(t_i, t_j) = min(|t_i(Month) - t_j(Month)|, min(t_i(Month), t_j(Month)) + 12 - max(t_i(Month), t_j(Month)))$$

$$t_i \theta_{Month} t_j \iff \Delta_{Month}(t_i(Month), t_j(Month)) \leq 4$$

Then θ_{Month} is the tolerance relation that considers two **tuples** similar if they have values within 4 months of distance, this is, their value in *Month* is similar.

A tolerance relation generates *blocks of tolerance*, among a set of tuples. A block of tolerance (w.r.t. an attribute a) is a group of tuples such that their values in attribute a are *similar*.

Definition 6.2 Given a set T, a subset $K \subseteq T$ and a tolerance relation $\theta \subseteq T \times T$. K is a *block of tolerance* of θ if:

1. $\forall t_i, t_j \in K : t_i \theta t_j$ (pairwise correspondence)
2. $\forall t_i \notin K, \exists t_j \in K : \neg(t_i \theta t_j)$ (maximality)

Given a set of tuples T and a set of attributes $\mathscr{U}$, for each attribute $x \in \mathscr{U}$, we define a tolerance relation θ_x on the values of x. The set of tolerance blocks generated by θ_x is denoted by T/θ_x.

Example 6.5 For example, when $T = \{1, 2, 3, 4, 5\}$, θ is defined as above, i.e. $v_1 \theta v_2 \iff |v_1 - v_2| \leq \varepsilon$, and $\varepsilon = 2$, then $T/\theta = \{\{1, 2, 3\}, \{2, 3, 4\}, \{3, 4, 5\}\}$.

We see that T/θ is not a partition, and this is because θ is not necessarily transitive. A partial ordering on the set of all possible tolerance relations in a set T can be defined as follows:

Definition 6.3 Let θ_1 and θ_2 two tolerance relations in the set T. We say that $\theta_1 \leq \theta_2$ if and only if $\forall K_i \in T/\theta_1 : \exists K_j \in T/\theta_2 : K_i \subseteq K_j$

This relation is a partial ordering and generates a lattice classifying tolerance relations, or, equivalently, blocks of tolerance. Given two tolerance relations, θ_1 and θ_2, the meet and the join operations in this lattice are:

Definition 6.4 Let θ_1 and θ_2 two tolerance relations in the set T.
$\theta_1 \wedge \theta_2 = \theta_1 \cap \theta_2 = max_T(\{K_i \cap K_j \mid K_i \in T/\theta_1, K_j \in T/\theta_2\})$
$\theta_1 \vee \theta_2 = \theta_1 \cup \theta_2 = max_T(T/\theta_1 \cup T/\theta_2)$
where $max_T(.)$ returns the set of maximal subsets w.r.t. inclusion.

Example 6.6 Based on Example 6.1, let us define the tolerance relation θ_m w.r.t. an attribute $m \in \{a,b,c,d\}$ as follows: $t_i \theta_m t_j \iff |t_i(m) - t_j(m)| \leq \varepsilon$. Then, assuming that $\varepsilon = 1$, we obtain:

$$T/\theta_a = \{\{t_1,t_3\},\{t_2,t_4\}\} \quad T/\theta_b = \{\{t_1,t_2,t_4\},\{t_3\}\}$$
$$T/\theta_c = \{\{t_1,t_2,t_3\},\{t_4\}\} \quad T/\theta_d = \{\{t_1,t_3\},\{t_2,t_4\}\}$$

We can check the following meet and join operations:

$$\theta_a \wedge \theta_b = \{\{t_1\},\{t_2,t_4\},\{t_3\}\} \quad \theta_a \vee \theta_b = \{\{t_1,t_2,t_3,t_4\}\}$$
$$\theta_a \wedge \theta_c = \{\{t_1,t_3\},\{t_2\},\{t_4\}\} \quad \theta_a \vee \theta_c = \{\{t_1,t_2,t_3,t_4\}\}$$
$$\theta_b \wedge \theta_c = \{\{t_1,t_2\},\{t_3\},\{t_4\}\} \quad \theta_b \vee \theta_c = \{\{t_1,t_2,t_3,t_4\}\}$$

We can also extend the definition of a similarity relation on a single attribute to a similarity relation to sets of attributes. Given $X \subseteq \mathcal{U}$, the similarity relation θ_X is defined as follows:

$$(t_i,t_j) \in \theta_X \iff \forall x \in X : (t_i,t_j) \in \theta_x$$

Two tuples are similar w.r.t. a set of attributes X if and only if they are similar w.r.t. *each* attribute in X.

6.3 FCA and Database Dependencies

In this section we present the two kinds of dependencies that will be discussed here: functional dependencies and similarity dependencies. In fact, the latter are a generalization of the former. As we will see, in order to switch from functional to similarity dependencies, we just need to change the implicit equivalence relation that is behind the definition of functional dependencies by a tolerance relation.

We also present the basics of FCA that are needed in order to understand some previous results concerning the characterization of FDs using FCA. We finish this section by presenting pattern structures, which is a generalization of FCA. We will use this formalism in the following section in order to present the main results in this article.

6.3.1 Functional Dependencies

Definition 6.5 ([35]) Let T be a set of tuples, and $X, Y \subseteq \mathcal{U}$. A **functional dependency (FD)** $X \rightarrow Y$ holds in T if:

$$\forall t, t' \in T : t(X) = t'(X) \implies t(Y) = t'(Y)$$

For instance, the functional dependencies $a \rightarrow d$ and $d \rightarrow a$ hold in Table 6.1, whereas the functional dependency $a \rightarrow c$ does not hold since $t_2(a) = t_4(a)$ but $t_2(c) \neq t_4(c)$.

6.3.2 Similarity Dependencies

Similarity Dependencies are a generalization or a relaxation of Functional Dependencies. The definition of similarity dependencies is the following:

Definition 6.6 Let $X, Y \subseteq \mathcal{U}$ and let θ be a tolerance relation: $X \rightarrow Y$ is a *similarity dependency* if: $\forall t_i, t_j \in T : t_i \theta_X t_j \implies t_i \theta_Y t_j$

While a functional dependency $X \rightarrow Y$ is based on equality of values, a similarity dependency $X \rightarrow Y$ holds if each pair of tuples having *similar* values w.r.t. attributes in X has *similar values* w.r.t. attributes in Y.

This definition just replaces the equality relation (an equivalence relation) for functional dependencies with a tolerance relation, which means that transitivity is dropped. In the following sections we will see that this generalization is captured by FCA in order to characterize both kinds of dependencies.

6.3.3 Formal Concept Analysis

Formal Concept Analysis (FCA) is a mathematical framework allowing to build a concept lattice from a binary relation between objects and their attributes. The concept lattice can be represented as a diagram where classes of objects/attributes and ordering relations between classes can be drawn, interpreted and used for data-mining, knowledge management and knowledge discovery [36].

We use standard definitions from [24]. Let G and M be arbitrary sets and $I \subseteq G \times M$ be an arbitrary binary relation between G and M. The triple (G, M, I) is called a formal context. Each $g \in G$ is interpreted as an object, each $m \in M$ is interpreted as an attribute. The statement $(g, m) \in I$ is interpreted as "g has attribute m". The two following derivation operators $(\cdot)'$:

$$A' = \{m \in M \mid \forall g \in A : gIm\} \quad for\ A \subseteq G,$$
$$B' = \{g \in G \mid \forall m \in B : gIm\} \quad for\ B \subseteq M$$

define a Galois connection between the powersets of G and M. The derivation operators $\{(\cdot)', (\cdot)'\}$ put in relation elements of the lattices $(\wp(G), \subseteq)$ of objects and $(\wp(M), \subseteq)$ of attributes and reciprocally. A Galois connection defines the closure operators $(\cdot)''$ and realizes a one-to-one correspondence between all closed sets of objects and all closed sets of attributes. For $A \subseteq G$, $B \subseteq M$, a pair (A, B) such that $A' = B$ and $B' = A$, is called a *formal concept*. Concepts are partially ordered by

$(A_1, B_1) \leq (A_2, B_2) \Leftrightarrow A_1 \subseteq A_2 \ (\Leftrightarrow B_2 \subseteq B_1)$. (A_1, B_1) is a sub-concept of (A_2, B_2), while the latter is a super-concept of (A_1, B_1). With respect to this partial order, the set of all formal concepts forms a complete lattice called the *concept lattice* of the formal context (G, M, I), i.e. any subset of concepts has both a supremum (join $\vee$) and an infimum (meet $\wedge$) [24]. For a concept (A, B) the set A is called the *extent* and the set B the *intent* of the concept. The set of all concepts of a formal context (G, M, I) is denoted by $\mathfrak{B}(G, M, I)$ while the concept lattice is denoted by $\underline{\mathfrak{B}}(G, M, I)$.

An implication of a formal context (G, M, I) is denoted by $X \rightarrow Y$, $X, Y \subseteq M$ and means that all objects from G having the attributes in X also have the attributes in Y, i.e. $X' \subseteq Y'$. Implications obey the Armstrong rules (reflexivity, augmentation, transitivity). The minimal subset of implications (in sense of its cardinality) from which all implications can be deduced with Armstrong rules is called the Duquenne-Guigues basis [25].

Objects described by non binary attributes can be represented in FCA as a many-valued context (G, M, W, I) with a set of objects G, a set of attributes M, a set of attribute values W and a ternary relation $I \subseteq G \times M \times W$. The statement $(g, m, w) \in I$, also written $g(m) = w$, means that "the value of attribute m taken by object g is w". The relation I verifies that $g(m) = w$ and $g(m) = v$ always implies $w = v$. For applying the FCA machinery, a many-valued context can be transformed into a formal context with a conceptual scaling. The choice of a scale should be wisely done w.r.t. data and goals since it affects the size, the interpretation, and the computation of the resulting concept lattice.

6.3.4 Functional Dependencies as Implications

In this section we present how functional dependencies are characterized with FCA (see [3] and [24]). This is performed by transforming a table into a formal context. The implications that hold in that formal context will be equivalent to the set of functional dependencies that hold in the original table.

Again, with a table D with attributes $\mathscr{U}$ taking values in *Dom*, we build the formal context $\mathbb{K} = (\mathscr{B}_2(G), M, I)$, where $G = T$ and $M = \mathscr{U}$ to respect the FCA notation from [24]. $\mathscr{B}_2(G) = \{ (t_i, t_j) \mid i < j \text{ and } t_i, t_j \in T \}$ is the set of all pairs of tuples from T (excluding symmetry and reflexivity to avoid redundancy) Then, the relation I is defined as

$$(t_i, t_j) \ I \ m \Leftrightarrow t_i(m) = t_j(m), \text{ for } m \in M$$

while attributes remain the same. Figure 6.1 illustrates the transformation of the initial data to build a formal context and its concept lattice.

The number of objects of this newly built formal context is in the range of $O(|T^2|)$ (where $|T|$ is the number of tuples), so it can be significantly larger than the original set of tuples T.

id	a	b	c	d
t_1	1	3	4	1
t_2	4	3	4	3
t_3	1	8	4	1
t_4	4	3	7	3

$\mathbb{K}$	a	b	c	d
(t_1,t_2)		×	×	
(t_1,t_3)	×		×	×
(t_1,t_4)		×		
(t_2,t_3)			×	
(t_2,t_4)	×	×		×
(t_3,t_4)				

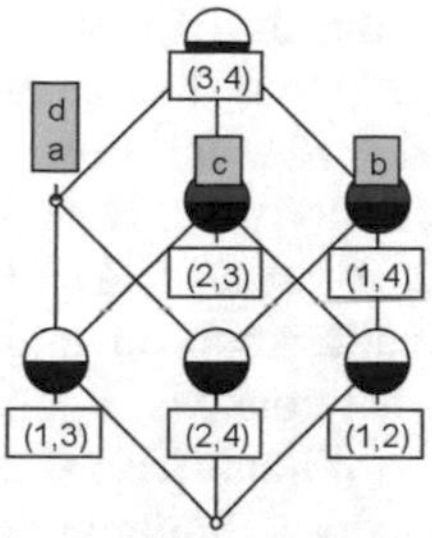

Fig. 6.1: Characterizing functional dependencies with FCA: from a set of tuples to a formal context and its concept lattice

We now explain how this concept lattice characterizes the set of all functional dependencies that hold in the table T with the following proposition:

Proposition 6.1 ([3, 24]) *A functional dependency $X \to Y$ holds in a table T if and only if $\{X\}'' = \{X,Y\}''$ in the formal context $\mathbb{K} = (\mathscr{B}_2(G), M, I)$.*

This proposition states how to test that a FD holds using the concept lattice that has been computed. For instance, let us suppose that we want to test whether a functional dependency $a \to b$ holds in the formal context of Fig. 6.1. We should test in the corresponding concept lattice if $\{a\}'' = \{a,b\}''$. In this particular case, we observe that $\{a\}'' = \{a,d\}$ and $\{a,b\}'' = \{a,b,d\}$, which means that this dependency does not hold in T. On the other hand, the dependency $ac \to d$ holds, since $\{a,c\}'' = \{a,c,d\}$ and $\{a,c,d\}'' = \{a,c,d\}$.

An interesting consequence is that the set of implications that hold in the formal context $\mathbb{K} = (\mathscr{B}_2(G), M, I)$ is syntactically equivalent to the set of functional dependencies that hold in a table T [3, 24]. By *syntactically* we mean that whenever an implication $X \to Y$ holds in $\mathbb{K}$, then the functional dependency $X \to Y$ holds in T.

6.3.5 Pattern Structures

A pattern structure is defined as a generalization of a formal context describing complex data [23]. Formally, let G be a set of objects, let $(D, \sqcap)$ be a meet-semi-lattice of potential object descriptions and let $\delta : G \longrightarrow D$ be a mapping associating each object with its description. Then $(G, (D, \sqcap), \delta)$ is a pattern structure. Elements of D are patterns and are ordered thanks to a subsumption relation $\sqsubseteq$: $\forall c, d \in D$, $c \sqsubseteq d \iff c \sqcap d = c$.

A pattern structure $(G, (D, \sqcap), \delta)$ is based on two derivation operators $(\cdot)^\square$:

$$d^\square = \{g \in G | d \sqsubseteq \delta(g)\} \quad for \ d \in (D, \sqcap).$$

These operators form a Galois connection between $(\wp(G), \subseteq)$ and $(D, \sqcap)$. Pattern concepts of $(G, (D, \sqcap), \delta)$ are pairs of the form (A, d), $A \subseteq G$, $d \in (D, \sqcap)$, such that $A^{\square} = d$ and $A = d^{\square}$. For a pattern concept (A, d), d is a pattern intent and is the common description of all objects in A, the pattern extent. When partially ordered by $(A_1, d_1) \leq (A_2, d_2) \Leftrightarrow A_1 \subseteq A_2$ ($\Leftrightarrow d_2 \sqsubseteq d_1$), the set of all concepts forms a complete lattice called pattern concept lattice.

As for formal contexts, implications can be defined. For $c, d \in D$, the pattern implication $c \to d$ holds if $c^{\square} \subseteq d^{\square}$, i.e. the pattern d occurs in an object description if the pattern c does. Similarly, for $A, B \subseteq G$, the object implication $A \to B$ holds if $A^{\square} \sqsubseteq B^{\square}$, meaning that all patterns that occur in all objects from the set A also occur in all patterns in the set B [23].

6.4 Results

In this section we show some of the most relevant results regarding the characterization of functional and similarity Dependencies with pattern structures. These results have appeared in [9, 11] for functional dependencies and in [8, 10] for Similarity Dependencies. We will see now that pattern structures are able to characterize functional and similarity dependencies in quite an elegant manner, without incurring in any extra transformation process.

6.4.1 Characterization of Functional Dependencies with Pattern Structures

The main idea behind this approach is to avoid the creation of a formal context $\mathbb{K} = (\mathscr{B}_2(G), M, I)$ (defined in Sect. 6.3.4) that, as it has already been discussed, has a size of order $O(|T^2|)$. The strategy consists in using Pattern Structures that will be built on the dataset, without any sort of transformation. Consider a dataset $D = (\mathscr{U}, T)$ as a many-valued context (G, M, W, J) where $G = T$ corresponds to the set of objects ("rows"), $M = \mathscr{U}$ to the set of attributes ("columns"), $W = Dom$ the data domain ("all distinct values of the table") and $J \subseteq G \times M \times W$ a relation such that $(g, m, w) \in J$ also written $m(g) = w$ means that attribute m takes the value w for the object g [24]. In the left table of Fig. 6.2, $b(t_4) = 3$.

We show how a partition pattern structure can be defined from a many-valued context (G, M, W, J) and show that its concept lattice is equivalent to the concept lattice of $\mathbb{K} = (\mathscr{B}_2(G), M, I)$ introduced above. Intuitively, formal objects of the pattern structure are the attributes of the many-valued context (G, M, W, J). Then, given an attribute $m \in M$, its description $\delta(m)$ is given by a partition over G such that any two elements g, h of the same class take the same values for the attribute m, i.e. $m(g) = m(h)$. The result is given in Fig. 6.2 (middle). As such, descriptions obey the ordering of a partition lattice as described above. It follows that (G, M, W, J) can be represented as a pattern structure $(M,(D,\sqcap),\delta)$ where M is the set of original attributes, and $(D,\sqcap)$ is the set of partitions over G provided with the partition intersection operation $\sqcap$. An example of concept formation is given as follows, starting from set $\{a,d\} \subseteq M$:

$$\{a,d\}^{\square} = \delta(a) \sqcap \delta(d)$$
$$= \{\{t_1,t_3\},\{t_2,t_4\}\} \sqcap \{\{t_1,t_3\},\{t_2,t_4\}\}$$
$$= \{\{t_1,t_3\},\{t_2,t_4\}\}$$
$$\{\{t_1,t_3\},\{t_2,t_4\}\}^{\square} = \{m \in M | \{\{t_1,t_3\},\{t_2,t_4\} \sqsubseteq \delta(m)\}$$
$$= \{a,d\}$$

Hence, $(\{a,d\},\{\{t_1,t_3\},\{t_2,t_4\}\})$ is a pattern concept. The resulting pattern concept lattice is given in Fig. 6.2 (right).

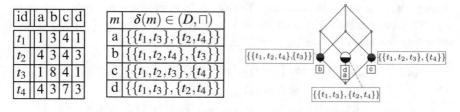

id	a	b	c	d
t_1	1	3	4	1
t_2	4	3	4	3
t_3	1	8	4	1
t_4	4	3	7	3

m	$\delta(m) \in (D,\sqcap)$
a	$\{\{t_1,t_3\},\{t_2,t_4\}\}$
b	$\{\{t_1,t_2,t_4\},\{t_3\}\}$
c	$\{\{t_1,t_2,t_3\},\{t_4\}\}$
d	$\{\{t_1,t_3\},\{t_2,t_4\}\}$

Fig. 6.2: The original data (left), the resulting pattern structure (middle) and its pattern concept lattice (right)

In the previous section, a many-valued context (G,M,W,J) was derived as the formal context $(\mathscr{B}_2(G),M,I)$ where $\mathscr{B}_2(G)$ represents any pair of objects, and $((g,h),m) \in I$ means that $m(g) = m(h)$. The resulting concept lattice is used to characterize the set of functional dependencies [24].

We also show that the lattice of pattern concepts yielded by the many-valued context (G,M,W,J) is isomorphic to the lattice of formal concepts that is yielded by the formal context $(\mathscr{B}_2(G),M,I)$:

Proposition 6.2 ([11]) (B,A) *is a pattern concept of the partition pattern structure* $(M,(D,\sqcap),\delta)$ *if and only if* (A,B) *is a formal concept of the formal context* $(\mathscr{B}_2(G),M,I)$ *for all* $B \subseteq M, A \subseteq \mathscr{B}_2(G)$ *(equivalently A is a partition on G).*

Following this equivalence, the following proposition shows that the functional dependencies that are computed in $(\mathscr{B}_2(G),M,I)$ and in (G,M,W,J) are exactly the same:

Proposition 6.3 ([11]) *A functional dependency* $X \to Y$ *holds in a table T if and only if:* $\{X\}^{\square} = \{XY\}^{\square}$ *in the partition pattern structure* $(M,(D,\sqcap),\delta)$.

Example. We consider a FD that holds in Table 6.1: $a \to d$. It is characterized from $(\mathscr{B}_2(G),M,I)$ as an attribute implication. It holds as well in $(M,\mathscr{B}_2(G),I)$ and $(M,(D,\sqcap),\delta)$ as an object implication.

6.4.2 Similarity Dependencies

In this section we present Proposition 6.4, the second main result of this article. As the reader will realize, the results in this section are very similar to those presented in the previous section. In fact, this is congruent with the fact that we are showing that the formalism of pattern structures is flexible enough to characterize not only dependencies that are based in an equivalence relation, but also with a *softer* relation, this is, a tolerance relation.

As in the previous case for functional dependencies, the dataset can be represented as a pattern structure $(\mathscr{U}, (D, \sqcap), \delta)$ where $\mathscr{U}$ is the set of original attributes. Here, $(D, \sqcap)$ is the set of sets of tolerance blocks (differently from the previous section, where we had a partition or equivalence relation) over the set of tuples T provided with the meet operation introduced in Definition 6.4. The description of an attribute $x \in \mathscr{U}$ is given by $\delta(x) = T/\theta_x$ which is given by the set of tolerance blocks.

Example 6.7 An example of concept formation in the pattern structure is given as follows. Consider the table in Example 6.1. Starting from the set $\{a,c\} \subseteq \mathscr{U}$ and assuming that $t_i \theta_x t_j \iff |t_i(x) - t_j(x)| \leq 2$ for all attributes:

$$\{a,c\}^{\square} = \delta(a) \sqcap \delta(c) = \{\{t_1,t_3\},\{t_2,t_4\}\} \sqcap \{\{t_1,t_2,t_3\},\{t_4\}\}$$
$$= \{\{t_1,t_3\},\{t_2\},\{t_4\}\}$$
$$\{\{t_1,t_3\},\{t_2\},\{t_4\}\}^{\square} = \{x \in \mathscr{U} \mid \{\{t_1,t_3\},\{t_2\},\{t_4\}\} \sqsubseteq \delta(x)\} = \{a,c\}$$

Hence, $(\{a,c\},\{\{t_1,t_3\},\{t_2\},\{t_4\}\})$ is a pattern concept.

Then the pattern concept lattice allows us to characterize all similarity dependencies holding in the set of attributes $\mathscr{U}$:

Proposition 6.4 ([10]) *An similarity dependency* $X \to Y$ *holds in a table* T *if and only if:* $\{X\}^{\square} = \{X,Y\}^{\square}$ *in the pattern structure* $(\mathscr{U}, (D, \sqcap), \delta)$.

This result is isomorphic to Proposition 6.3. That means that the softening of the relation between tuples (equality in the case of functional dependencies, similarity in this present case) does not prevent the pattern structure from characterizing these dependencies. This is due in part to the fact that a partial order can also be defined over tolerance blocks. We do not mean that this condition is sufficient, but it is evidently necessary.

6.5 Discussion

We have revisited some results that concern the characterization of two different kinds of dependencies with pattern structures. What is the interest of using this formalism for the computation of dependencies that have already been computed using different methods not related to lattice theory? The answer is many-fold:

1. **A unified framework**. We have seen that two different kinds of dependencies may be handled using the same formalism (pattern structures). Obviously, the parameters for this formalism (objects, attributes, domain, relation) depend on the kind of dependency that we are dealing with, but as we have seen in Propositions 6.3 and 6.4, the characterization of both kinds of dependencies are equivalent.

   This is also true for other different kinds of dependencies: order-like dependencies [9, 18], acyclic join dependencies [5], degenerate multivalued dependencies [6], to name some of them. If we examine all these results, FCA offers an **encapsulation** of the semantics of each kind of dependency so that their syntax is handled by the pattern structures.

2. **Common algorithms**. As a consequence of the previous remark, we observe that potentially the same algorithms that have been used to compute formal concepts, implications, concept lattices, minimal (Duquenne-Guigues) bases can be used for all characterizations. One of the most remarkable examples is the computation of the implications that hold in a (transformed) context in Sect. 6.3.4, which are, in fact, the functional dependencies that hold in a dataset.

3. **New semantics**. Because we are using FCA and pattern structures in order to characterize dependencies, we also have structures that were unknown in database theory: formal concepts, concept lattices, etc. These objects have not been yet explored in the scenarios that we have described in this paper. Said otherwise: how can a concept lattice help a database practitioner?

4. **Different classification**. In [1, 30] there are different criteria in order to classify dependencies in the relational database model (tuple or non-tuple generating, typed or non-types, etc). However, with FCA there is no difference between the characterization of tuple-generating dependencies like multivalued dependencies [3, 7] and non tuple-generating dependencies like functional dependencies [11]. In fact, the difference is encapsulated within the relation between objects and attributes, and the difference shows up the different kinds of attributes of the formal context: partitions of the attribute set in the former case, sets of attributes in the latter. Could we devise a different classification of dependencies w.r.t. FCA?

We suggest some items for discussion that may be of interest to our FCA community as well as the database community that, eventually, may become interested in our research:

1. **Solve problems that are present in the database community**. To name just a few of them: dependencies with complex data (graphs, sequences, etc. . .), storage limitations, fast computation of queries.

2. **Do not fight the standard**. We need to admit that a standard in a long-time established community with a large number of users like that of the database community is a difficult thing to change. Even if that standard is too complex, obscure or anti-intuitive w.r.t. other formulations, the fact that this formalism has survived for such a long time in such a huge community means that it will stay there. For instance, if we take the definitions of a closure in basic texts of

database theory [1, 30], we see that the equivalent in FCA is much clearer and easy to understand (maybe, this is a biased point of view of someone who has been working with FCA for years). But even if this is the case, that community will not change their formalism by our new proposal. Therefore, instead of offering them a completely new formalism, our task should be to offer them different and *more efficient* solutions to their problems, not expecting that they would necessarily embrace FCA as a new paradigm.

3. **Have the practitioner in mind**. It is clear that some victories achieved by FCA outside our community have to do with applications rather than theory. Obviously, theory is the *sine qua non* condition for the practical solutions, but, again, what seems appealing to people outside our community is the potential use of FCA as a problem solving tool. In the case of the database community, we think that this is also true: they already have their formalisms, which are very unlikely to be changed. But they have a set of open problems that may be solved by FCA or some old problems which may be better solved using FCA.

Obviously, we just mention these points as a departure point for debate. We do not assume that all the previous points need to be an absolute truth, we just hope that they can be of interest for researchers in our community.

6.6 Conclusions

We have seen that FCA and pattern structures can characterize different kinds of dependencies in an elegant and compact way. These results are also present for different kinds of database dependencies. We have shown the advantages and some potential drawbacks that FCA offers to the database community.

We think that there has been an enormous amount of work in this fields that needs to be continued. The results presented in this paper show that there are still new lines of research that need to be explored, and that may be fruitful not only to the FCA community but, most importantly, to the database community.

Acknowledgements This research was supported by the recognition of 2017SGR-856 (MACDA) from AGAUR (Generalitat de Catalunya), and the grant TIN2017-89244-R from MINECO (Ministerio de Economía y Competitividad).

References

1. S. Abiteboul, R. Hull, and V. Vianu. *Foundations of Databases*. Addison-Wesley, Reading (MA), USA, 1995.
2. J. Baixeries. A formal concept analysis framework to model functional dependencies. In *Mathematical Methods for Learning*, 2004.

3. J. Baixeries. *Lattice Characterization of Armstrong and Symmetric Dependencies (PhD Thesis)*. Universitat Politècnica de Catalunya, 2007.
4. J. Baixeries. *A Formal Context for Symmetric Dependencies*, pages 90–105. Springer Berlin Heidelberg, Berlin, Heidelberg, 2008.
5. J. Baixeries. A formal context for acyclic join dependencies. In M. Kryszkiewicz, A. Appice, D. Ślezak, H. Rybinski, A. Skowron, and Z. W. Raś, editors, *Foundations of Intelligent Systems*, pages 563–572, Cham, 2017. Springer International Publishing.
6. J. Baixeries and J. L. Balcázar. Characterization and armstrong relations for degenerate multivalued dependencies using formal concept analysis. In B. Ganter and R. Godin, editors, *ICFCA*, volume 3403 of *Lecture Notes in Computer Science*, pages 162–175. Springer, 2005.
7. J. Baixeries and J. L. Balcázar. A lattice representation of relations, multivalued dependencies and armstrong relations. In *ICCS*, pages 13–26, 2005.
8. J. Baixeries, V. Codocedo, M. Kaytoue, and A. Napoli. Characterizing approximate-matching dependencies in formal concept analysis with pattern structures. *Discrete Applied Mathematics*, 249:18 – 27, 2018. Concept Lattices and Applications: Recent Advances and New Opportunities.
9. J. Baixeries, M. Kaytoue, and A. Napoli. Computing functional dependencies with pattern structures. In L. Szathmary and U. Priss, editors, *CLA*, volume 972 of *CEUR Workshop Proceedings*, pages 175–186. CEUR-WS.org, 2012.
10. J. Baixeries, M. Kaytoue, and A. Napoli. Computing similarity dependencies with pattern structures. In M. Ojeda-Aciego and J. Outrata, editors, *CLA*, volume 1062 of *CEUR Workshop Proceedings*, pages 33–44. CEUR-WS.org, 2013.
11. J. Baixeries, M. Kaytoue, and A. Napoli. Characterizing Functional Dependencies in Formal Concept Analysis with Pattern Structures. *Annals of Mathematics and Artificial Intelligence*, 72(1–2):129–149, Oct. 2014.
12. R. Belohlávek and V. Vychodil. Data tables with similarity relations: Functional dependencies, complete rules and non-redundant bases. In M.-L. Lee, K.-L. Tan, and V. Wuwongse, editors, *DASFAA*, volume 3882 of *Lecture Notes in Computer Science*, pages 644–658. Springer, 2006.
13. M. Baudinet, J. Chomicki, and P. Wolper. Constraint-generating dependencies. *J. Comput. Syst. Sci.*, 59(1):94–115, 1999.
14. L. Bertossi, S. Kolahi, and L. V. S. Lakshmanan. Data cleaning and query answering with matching dependencies and matching functions. In *Proceedings of the 14th International Conference on Database Theory*, ICDT '11, pages 268–279, New York, NY, USA, 2011. ACM.
15. P. Bohannon, W. Fan, F. Geerts, X. Jia, and A. Kementsietsidis. Conditional functional dependencies for data cleaning. In *ICDE*, pages 746–755, 2007.
16. L. Caruccio, V. Deufemia, and G. Polese. Relaxed functional dependencies - A survey of approaches. *IEEE Trans. Knowl. Data Eng.*, 28(1):147–165, 2016.
17. N. Caspard and B. Monjardet. The lattices of closure systems, closure operators, and implicational systems on a finite set: A survey. *Discrete Applied Mathematics*, 127(2):241–269, 2003.
18. V. Codocedo, J. Baixeries, M. Kaytoue, and A. Napoli. Characterization of Order-like Dependencies with Formal Concept Analysis. In M. Huchard and S. Kuznetsov, editors, *Proceedings of the Thirteenth International Conference on Concept Lattices and Their Applications, Moscow, Russia, July 18–22, 2016.*, volume 1624 of *CEUR Workshop Proceedings*, pages 123–134. CEUR-WS.org, 2016.
19. V. Codocedo, J. Baixeries, M. Kaytoue, and A. Napoli. Contributions to the Formalization of Order-like Dependencies using FCA. In *What can FCA do for Artificial Intelligence?*, The Hague, Netherlands, Aug. 2016.
20. J. Demetrovics, G. Hencsey, L. Libkin, and I. B. Muchnik. Normal form relation schemes: A new characterization. *Acta Cybern.*, 10(3):141–153, 1992.

21. W. Fan, H. Gao, X. Jia, J. Li, and S. Ma. Dynamic constraints for record matching. *The VLDB Journal*, 20(4):495–520, Aug. 2011.

22. A. Floratou, N. Teletia, D. J. DeWitt, J. M. Patel, and D. Zhang. Can the elephants handle the nosql onslaught? *Proc. VLDB Endow.*, 5(12):1712–1723, Aug. 2012.

23. B. Ganter and S. O. Kuznetsov. Pattern Structures and Their Projections. In *Proceedings of ICCS 2001*, Lecture Notes in Computer Science 2120, pages 129–142. Springer, 2001.

24. B. Ganter and R. Wille. *Formal Concept Analysis*. Springer, Berlin, 1999.

25. J.-L. Guigues and V. Duquenne. Familles minimales d'implications informatives résultant d'un tableau de données binaires. *Mathématiques et Sciences Humaines*, 95:5–18, 1986.

26. M. Kaytoue, S. O. Kuznetsov, A. Napoli, and S. Duplessis. Mining gene expression data with pattern structures in Formal Concept Analysis. *Information Sciences*, 181(10):1989–2001, 2011.

27. S. O. Kuznetsov. Machine learning on the basis of formal concept analysis. *Autom. Remote Control*, 62(10):1543–1564, Oct. 2001.

28. S. O. Kuznetsov. Pattern Structures for Analyzing Complex Data. In H. Sakai, M. K. Chakraborty, A. E. Hassanien, D. Slezak, and W. Zhu, editors, *RSFDGrC*, volume 5908 of *Lecture Notes in Computer Science*, pages 33–44. Springer, 2009.

29. S. Lopes, J.-M. Petit, and L. Lakhal. Functional and approximate dependency mining: database and fca points of view. *Journal of Experimental and Theoretical Artificial Intelligence*, 14(2–3):93–114, 2002.

30. D. Maier. *The Theory of Relational Databases*. Computer Science Press, 1983.

31. H. Mannila and K.-J. Räihä. *The Design of Relational Databases*. Addison-Wesley, Reading (MA), USA, 1992.

32. R. Medina and L. Nourine. Conditional functional dependencies: An fca point of view. In L. Kwuida and B. Sertkaya, editors, *ICFCA*, volume 5986 of *Lecture Notes in Computer Science*, pages 161–176. Springer, 2010.

33. S. Song and L. Chen. Efficient discovery of similarity constraints for matching dependencies. *Data & Knowledge Engineering*, 87(0):146 – 166, 2013.

34. S. Song, L. Chen, and P. S. Yu. Comparable dependencies over heterogeneous data. *The VLDB Journal*, 22(2):253–274, Apr. 2013.

35. J. Ullman. *Principles of Database Systems and Knowledge-Based Systems, volumes 1–2*. Computer Science Press, Rockville (MD), USA, 1989.

36. R. Wille. Why can concept lattices support knowledge discovery in databases? *Journal of Experimental and Theoretical Artificial Intelligence*, 14(2–3):81–92, 2002.

Chapter 7
Leveraging Closed Patterns and Formal Concept Analysis for Enhanced Microblogs Retrieval

Meryem Bendella and Mohamed Quafafou

7.1 Introduction

Social microblogging services offered by the web 2.0 have gained a significant interest for society during our decade. Twitter[1] is one of the most popular social microblogging platforms that enable users to post short texts (up to 280 characters for one text) called tweets. Nowadays, users can share ideas, opinions, emotions, suggestions, daily stories and events through this platform. Many of these online services start to be part of daily life of millions of people around the world. However, a huge quantity of information is created in these platforms, hence, finding recent and relevant information is challenging.

In addition to publishing messages (microblogs), there are many users who are interested in collecting recent information from such platforms. This information can be related to a particular event, a specific topic or popular trends. Users express their need through a query to search posts (tweets). According to [35], most people use few search terms and few modified queries in Web searching. However, query formulation becomes difficult for users in order to express appropriately what they are looking for. Query Expansion (QE) plays a considerable contribution towards fetching relevant results in this case. An expanded query will contain more related terms (called candidate terms) to increase the chances of rendering the maximum number of relevant documents. The objective is to find the microblogs answering to a need for information specified by a user.

In this work, we propose a new method based on Formal Concept Analysis (FCA) and word embeddings to expand user queries. In order to achieve this goal, we prepare our dataset collection by preprocessing the tweet text which is a critical step in information retrieval (IR) and Natural Language Processing (NLP). Then, each tweet is represented as a set of words and will be indexed by the Terrier

---

[1] https://twitter.com/.

M. Bendella (✉) • M. Quafafou
Aix-Marseille Université, Marseille, France
e-mail: meryem.bendella.billami@gmail.com; mohamed.quafafou@univ-amu.fr

© The Author(s), under exclusive license to Springer Nature Switzerland AG 2022
R. Missaoui et al. (eds.), *Complex Data Analytics with Formal Concept Analysis*,
https://doi.org/10.1007/978-3-030-93278-7_7

system.[2] Next, we use this system to retrieve tweets according to the TREC2011 query set [27]. We used the BM25 [17] model to retrieve relevant tweets answering original queries. After that, these retrieved tweets are used to extract frequent closed patterns. This can be defined as the frequent closed patterns of words contained in tweets dataset. Furthermore, word embeddings are trained on our textual dataset by using Word2Vec model [24]. Indeed, we expand original queries by combining patterns and word embeddings approaches. This combination consists of enriching the query by finding most closely related words to the words of the pattern. The proposed model extends the semantics used in the original query and improves the results search of microblogs.

The paper is structured as follows. We cover related work on query expansion for microblogs retrieval in Sect. 7.2. We discuss the FCA-based query expansion approach in Sect. 7.3. Then, Sect. 7.4 describes the proposed approach for expanding queries in microblogs retrieval. The Sect. 7.5 is dedicated to experiments and evaluations. Finally, the conclusion and future work are presented in Sect. 7.6.

7.2 Related Work

Query expansion (QE) has recently gained growing attention in IR domain and there is considerable research addressing the short query problem. Web queries posted by users can be too short and that makes the search results not focused on the topic of interest.

Much effort has been made to improve microblog retrieval. Reference [20] explored use of three different IR query expansion techniques in order to enhance the search results. In [23] authors propose a retrieval model for searching microblog posts for a given topic of interest. This model is based on a combination of quality indicators and the query expansion model.

Query expansion approaches for microblog retrieval can be divided into three groups which are local, global and external [28]:

- **Local**: Local QE techniques select candidate expansion terms from a set of documents retrieved in response to the original (unexpanded) query. This kind of approach is known as Pseudo-Relevance Feedback (PRF). It is widely used for query expansion in research of microblog search [10, 23, 39]. This approach consists in using terms derived from the top N retrieved documents (relevant documents) to retrieve other similar documents which are also likely to be relevant. In [19], authors propose an algorithm of extracting features from tweets using frequent patterns. Their QE method is based on PRF approach by applying weights for different features. They conducted experiments on TREC Microblog data (*cf.* Sect. 7.5.1).

---

[2] Terrier is an effective open source search engine (Information Retrieval system), readily deployable on large-scale collections of documents.

There local query expansion approaches have some limitations. In case of microblogs retrieval, where queries are short and ambiguous, the first documents retrieved by the system are not necessarily relevant. This can lead to create candidate terms which are unrelated to the initial query. These limits can have a negative impact on the results [22].

- **Global**: Global QE approaches select the expansion terms from the entire database of documents. These techniques select candidate terms by mining term-term relationships from the target corpus [28]. Global analysis techniques provide better results compared to techniques based on the Pseudo Feedback of Relevance (local analysis) [36]. This type of techniques has been used in many applications [15, 25], and was one of the first techniques to produce consistent effectiveness improvements through automatic expansion [7]. In [6], the proposed approach consisted in the use of term relationships in query expansion within the Language Model (LM) framework. They also integrated relationships computed by information flow into a LM. Authors showed that the idea of query expansion with term relationships can be naturally implemented in LM. In [13], authors used the semantic network WORDNET for developing a thesauri based on word co-occurrences.
  In our work, we use, in a certain way, the global analysis by training word embeddings on the entire dataset in order to extract terms that are most similar to the patterns. In [37], authors propose an QE method using word embeddings and some external resources.

- **External**: External QE techniques comprise methods that obtain expansion terms from other resources besides the target corpus [28]. Several approaches have been proposed to use external resources such as *Wikipedia*, *WordNet* and *DBpedia* to improve query expansion [1, 18, 40]. In [1], authors proposed an approach for expanding queries using external resources. Their approach includes the generation of candidate concepts from *Wikipedia* and *DBpedia*, as well as the selection of K-best concepts according to the scores of explicit semantic analysis. Also, *Wikipedia* was used by ALMasri et al. [3] to retrieve semantically related terms from the initial query for the purpose of enriching it. Furthermore, authors in [40] propose a query expansion method based on association rules extraction and external sources to extend short queries.

In addition, hybrid expansion approaches have also been proposed for the purpose of combining two or more expansion methods. In [41], authors proposed a hybrid model which is basically a combination of external resources with association rules. This model aims to improve the process of generating and selecting expansion terms (also called candidate terms).

7.3 FCA-Based Query Expansion

Frequent pattern has an important and active role in many data mining tasks [14]. This comprises to find interesting patterns (sets of items) called frequent itemsets from databases. It was initiated by Agrawal et al. [2] and it corresponds to finding the sets of attributes (or items) that appear simultaneously in at least a certain number of objects (or transactions) defined in an extraction context. The transactional dataset is represented by a set of preprocessed tweets. Each tweet is represented as a set of words which are considered as itemsets.

In this work, we are interested in frequent closed patterns which has been proposed by [29]. The pattern X is called closed if none of its supersets have the same support as X. In other words, $\forall Y,\ X \subseteq Y,\ support(Y) < support(X)$. Here, Y is a superset of X. Formal concept analysis (FCA) is a theory of data analysis identifying the conceptual structures within data sets. It also presents an interesting unified framework to identify dependencies among data, by understanding and computing them in a formal way [9]. This gives it the advantage to be an effective technique to analyze the different pattern relationships such as in Social Network [34], and Information Retrieval [8]. Given a formal context $\mathscr{D}$, there is a unique ordered set which describes the inherent lattice structure defining natural groupings and relationships among the transactions and their related items. This structure is known as a concept lattice or Galois lattice [12]. Each element of the lattice is a pair (T, I) which consists of a set of transactions and a set of items.

Let X be a closed pattern of items (words), a formal concept is composed of X and of the set of tweets containing this closed pattern.

7.3.1 Patterns Discovery

We perform patterns extraction on top-N tweets returned by Terrier system in the initial search. The topmost relevant tweets are retrieved by using Terrier with original queries. After that, we discover the closed frequent itemsets in this large database of transactions (retrieved tweets). This process is performed by using Charm-L algorithm [38], where the discovery of patterns computes the closed sets of items (i.e., words) that appear together in at least a certain number of transactions (i.e., tweets) recorded in a database. This number is called threshold and it is defined empirically. We compute the frequent concept lattice $\mathscr{L}$ according to the minimal support threshold value minsup. Each node of the formal concept lattice $\mathscr{L}$ represents the correspondence between a pattern and the set of tweets that contain the words of this pattern. The parameters N (for N-top tweets) and *minsup* are chosen according to experiments we conducted (described in Sect. 7.5.4).

The algorithm for generating the complete set of interesting frequent closed patterns and selection of candidate terms for the query expansion is shown in Algorithm 1, with L_K denoting the set of K closed frequent itemsets, T the set of N transactions which represent tweets, *minsup* the minimal support threshold value, and $\mathcal{Q}_P$ selected patterns for the expanded query. The list of K patterns contains all interesting patterns found in top-N retrieved documents according to the initial query.

From the list of K frequent concepts (frequent closed itemsets and transactions) found by applying the Algorithm 1, we select K patterns to represent the candidate

---

Algorithm 1 Patterns extraction

Require:
    $\mathcal{Q}$: initial query
    $\mathcal{T}$: a set of N tweets corresponding to $\mathcal{Q}$
    $\mathcal{W}$: Vocabulary of all words contained in tweets dataset
    minsup: Minimum support threshold value
    $\mathcal{R}$: a binary relation where $\mathcal{R} \subseteq \mathcal{T} \times \mathcal{W}$
Ensure:
    $\mathcal{L}_K$: List of K patterns

1: Creation of formal context $\mathcal{D} = (\mathcal{T}, \mathcal{W}, \mathcal{R})$;
    /* Computation of frequent concept lattice $\mathcal{L}$ for $\mathcal{D}$ according to *minsup* */
2: $\mathcal{L}_K = CharmL(\mathcal{T}, minsup)$;
3: $\mathcal{L}_{K_i} = argmax(Support(\mathcal{L}_{K_i}))$
4: **return** $\mathcal{L}_K$;

---

terms for expanding queries. This selection is based on the support of each pattern (frequent concept transactions). The list of frequent concepts is sorted in ascending order according to the value of the support (number of transactions in the concept). Then, among the list of the first patterns, the formal concept analysis allows us to select the terms that are related and similar. The terms of the selected patterns are then chosen as candidate terms (K patterns). For example, given a query $\mathcal{Q}$, we have $\mathcal{L}_{K_i}$ patterns. Let $\mathcal{Q}_P$ be the set of candidate terms. This set is obtained from the K patterns where $\mathcal{L}_{K_i} = argmax(Support(\mathcal{L}_{K_i}))$. Among these K patterns, we select terms that we use to extend the initial query based on formal concept analysis. This allows the identification of groups of transactions (tweets) having items (terms) in common. We detail in the following section how these patterns are extended by word embedding.

7.4 Patterns and Word Embeddings Based Query Expansion

Figure 7.1 gives an overview of our proposed method to expand queries for the microblog retrieval task. The proposed method is based on frequent concept extraction and Word Embeddings. It is composed of three main steps: (1) Generate and select patterns (see Sect. 7.3), (2) Extend patterns using Word Embeddings, and (3) expand the query.

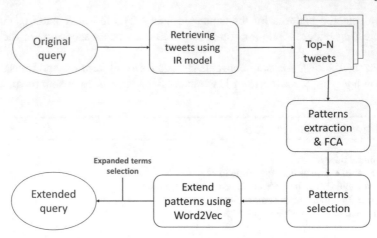

Fig. 7.1: Overview of the proposed query expansion approach

7.4.1 Word Embeddings: Word2Vec Model

Word Embeddings has been used in query expansion task to enhance search [32]. The main idea of word embeddings is to project words in a space in which semantic relations between these words can be observed. The words are represented in a continuous vector space of dimension n. An embedding can be seen as a function $M : \{V \rightarrow \mathbb{R}^n : w \mapsto \mathbf{w}\}$, which consists of projecting the word w from the vocabulary V into n-dimensional continuous space $\mathbb{R}$ ($\mathbb{R}^n$).

Word embeddings have already been successfully applied to the expansion of queries to improve Information Retrieval (IR) [4, 32]. Different approaches using word embedding have been proposed to expand queries in IR. The main difference lies in the way these techniques are used for extracting and selecting similar words to enrich the query.

Our approach uses these techniques by combining them with patterns to extend queries in microblogs (tweets) search . We relied on the corpus of documents used for retrieval as a learning set for embeddings. In [4] and [10], authors found that using the IR corpus for learning generally produces better results. In [11] authors conducted some experiments in order to measure the impact of learning dataset for training word embeddings. The results obtained by their study showed that it is preferable to train embeddings on the corpus used by the IR system.

In this work, we combine word embeddings with frequent closed patterns to expand queries for microblogs retrieval. We train word embeddings on the entire dataset in order to extract terms that are most similar to the selected patterns computed in the previous step. For each query, we extend terms of the selected patterns by adding the most similar terms contained in the dataset that are likely to be relevant to the query but do not appear in the patterns. For the training of *embeddings*, we used a project alternative WORD2VEC[3] implemented in programming language JAVA by MEDALLIA team[4] to integrate it into our main program of query expansion implemented in JAVA. The training of the neural network is carried out on the preprocessed corpus TREC 2011 on which search is performed. This model estimates the probability that a term will appear in a position in a text based on terms that appear in a window around this position. Each term in the dataset is represented by a vector embedded in a vector space. Similarities between these vectors were shown to correspond to semantic similarities between terms [24].

For the setting of the neural network, we have set a window with a size of 7 words (the appearance frequency of the words is at least 5, the dimensions of the vectors are 200, the number negative examples is 7 with a use of the hierarchical alternative *softmax*. Specifically, the continuous bag of words model (CBOW) is used.

7.4.2 Expansion Terms Selection

Given an original query $q = \{w_{q1}, \ldots, w_{qn}\}$, the process of expanding q is threefold: (1) retrieving relevant tweets answering q using a retrieval model, (2) selecting a set of candidate terms (CT) for q by extracting *patterns* from the top-N ranked tweets, and using formal concept analysis, (3) selecting the most related terms to the CT set using *Word2Vec* model, in order to add only terms that are semantically related to q. These terms are then selected to obtain the set of terms which represent the expanded query denoted eq, with $eq = q \bigcup \{w_{eq1}, \ldots, w_{eqm}\}$.

The process of obtaining candidate terms consists in selecting terms from the patterns set. For extending these terms, we used Word2Vec model in order to select terms that are semantically related to the obtained patterns. The process for selecting these terms computes the cosine similarity between the corresponding vector of pattern terms and each vector of tweet terms in the corpus, and rank the words in decreasing order of the cosine similarity.

We give in Table 7.1 an example of query expansion by applying the proposed approach. The example of query is resulting from TREC 2011 collection. The first three words of the expanded query represent the candidate terms obtained by patterns extraction. Thus, the other words represent the words obtained by applying the similarity measure (word embeddings). The final query is therefore given by merging the initial query with the expanded query.

Initial query	Expanded query	Final query
Mexico drug war	Mexico border violence Drug smuggler spill Cannabi cancun	Mexico drug war border violence smuggler spill cannabi cancun

Table 7.1: Example of expanded query using Patterns and Word Embeddings

---

7.5 Experiments

In this section, we conduct experiments to evaluate the effectiveness of the proposed query expansion method. To demonstrate the performance of our proposed method, we compare our patterns-based query expansion method with several methods. Our experiments are conducted on the TREC 2011 collection.

7.5.1 Dataset Description

For our experiments, we initially used a collection of tweets collected via Twitter APIs. This collection is used to choose the parameters of our method which are the number of frequent patterns and the number of terms selected by the cosine similarity measure. Furthermore, we used TREC data collection proposed for TREC 2011 Microblog Track Data, in order to evaluate the proposed approach. This dataset contains approximately 16 million tweets collected over a period of 2 weeks (24th January 2011 until 8th February) [27]. Since the provided dataset contains only tweet ids, on the whole, we have gathered around 12 million tweets with content. For the rest of the collection, tweets either was removed by their editors or we have no access to them. After performing the data filtering and processing task explained below, we obtained a dataset of around 3.5 million tweets on which our experiments were conducted. We have used 49 queries defined by TREC track 2011 [27]. An example of a query with the TREC format is described as follows:

```
<top>
    <num> Number: MB001 </num>
    <title> BBC World Service staff cuts </title>
    <querytime> Tue Feb 08 12:30:27 +0000 2011 </querytime>
    <querytweettime> 34952194402811904 </querytweettime>
</top>
```

We performed a preprocessing of the text of the tweet i.e., dealing with stop words, emoticons, punctuation, stemming, etc. We remove stopwords and emoticons from each tweet. Microblogs data such as tweets are too short, generally not well written and do not respect the grammar. However, this preliminary step is very crucial to eliminating the noise and cleaning data. We, therefore, prepare the dataset for indexing by filtering tweets as follows: (1) Removing null tweets and short tweets which contain less than two words; (2) Removing Retweets (tweets starting with RT followed by username) as they would be judged as non-relevant; (3) Removing non-English tweets; (4) Eliminating the non-ASCII content found in any of the English tweets; (5) Removing link and mentions from the tweet.

Furthermore, we perform tokenization, tweet normalization, text stemming, and stopwords removal, as part of the preprocessing phase. Tokenization is the process of breaking each tweet up into words or other meaningful elements called tokens. Then, we stem all tokens present in tweets except hashtags. We have used the standard Porter stemmer of Stanford NLP tool.[5] After that, we remove English stopwords[6] that are present in tweets. We also perform normalization of the tweet content, by resolving words containing many repeated letters, such as the word "yes" or "happy", they may appear as "yeeees" or "happyyy" on Twitter. The output of preprocessing task is used for indexing.

7.5.2 Retrieval Model

We use the well-known Terrier IR system[7] to index our data collection (TREC 2011). Terrier, an open source software, offers a range of document weighting and query expansion models. It has been successfully used for ad-hoc retrieval, cross-language retrieval, Web IR and intranet search [26]. All the tweets (after the preprocessing task) were indexed using Terrier, and the original queries were used to retrieve and rank tweets using the standard BM25[8] retrieval model [17, 30] of Terrier. BM25 model has been used extensively within the TREC community on a variety of corpora. The function S_{BM25} described in the Eq. 7.1 returns the weight computation score for the BM25 model.

$$S_{BM25} = \sum_{t_i \in T} \frac{(k_1 + 1) \cdot tf_i}{k_1 \cdot ((1 - b) + b \cdot \frac{dl}{avgdl}) + tf_i} \cdot log\left(\frac{N - R_i}{R_i}\right) \qquad (7.1)$$

tf_i depicts t_i term frequency in document. dl and $avgdl$ represent the length of document and the average length of document in the collection respectively. N is the number of documents for all the collection and R_i denotes the number of documents containing the term t_i. k_1 and b are the model parameters.

7.5.3 Experimental Protocol

We use Precision, MAP and nDCG metrics, which are widely used in information retrieval [33], to evaluate the proposed method for query expansion. Moreover, we evaluated performance of the proposed method according to its Precision, Recall, F-measure, and R-PREC metrics in order to compare the performance and effectiveness

---

[5] https://nlp.stanford.edu/IR-book/html/htmledition/stemming-and-lemmatization-1.html.

[6] https://github.com/ravikiranj/twitter-sentiment-analyzer/blob/master/data/feature_list/stopwords.txt.

[7] http://terrier.org/.

[8] http://terrier.org/docs/v4.0/javadoc/org/terrier/matching/models/BM25.html.

of the proposed approach with other approaches. R-PREC is the precision after R documents retrieved. The MAP (Mean Average Precision) for a set of queries is the mean of the average precision scores for each query. MAP is computed as follows:

$$MAP = \frac{\sum_{q=1}^{Q} AveP(q)}{Q} \qquad (7.2)$$

Normalized discounted cumulative gain (nDCG) is a measure of retrieval quality for ranked documents that, in contrast to precision, makes use of graded relevance assessments [16]. nDCG is computed as follows:

$$nDCG = Z_i \sum_{j=1}^{R} \frac{2^{r(j)} - 1}{log(1+j)'} \qquad (7.3)$$

Here, Z_i is a constant to normalize the result to the value of 1. $r(j)$ is an integer representing the relevance level of the result returned at rank j where R is the last possible ranking position. In our case, the relevance levels are 0 (irrelevant), 1 (relevant), and 2 (high relevant). nDCG@n is a variation of nDCG where only the top-n results are considered.

For our experimental evaluation, we compute nDCG@10, MAP, P@5, P@10, and P@30 by using TREC EVAL.[9] P@5, P@10 and P@30 represent respectively, precision considering only the top 5, top 10, and top 30 results returned by the system.

7.5.4 Experimental Results

We evaluate the effectiveness of our proposed approach by using evaluation metrics detailed in Sect. 7.5.3. We conducted two different runs in our experiments, the first one is based on patterns, and the second one represents the combination of patterns and Word Embeddings.

Table 7.2 reports the performance of these runs compared with the baseline run. In this work, we define our baseline as a single run which was generated using Terrier system, selecting the most recent 1000 tweets that contain any of the query terms. In other words, without any query expansion process. We also carried out a test on a method based on Pseudo-Relevance Feedback (PRF) in order to compare our results. This expansion method is applied using the *Bo1* term weighting model implemented in terrier [5]. It is a statistic-based expansion model which works similarly to the *Rocchio* model [31]. We took into consideration the default configuration proposed by the system (the number of N first documents : 3; the number of terms used to expand the query : 10).

---

[9] https://trec.nist.gov/trec_eval/.

Metrics	P@10	P@30	MAP	nDCG@10	nDCG
Baseline	0.1184	0.0905	0.1025	0.1146	0.2659
PRF	0.2245	0.2116	0.1759	0.2067	0.3876
Run-P	0.2980	0.2415	0.1929	0.2619	0.3938
Run-P-WE	0.3449	0.2878	0.2403	0.3077	0.4476

Table 7.2: Evaluation results of the proposed approach

Compared to the *baseline*, we have achieved some major improvements on the four measures: +191,30%, +218,01%, +134,43% et +168,5% respectively on P@10, P@30, MAP and nDCG@10. Therefore, the proposed method gives a precision (*P@10*) of 0.3449 versus 0.2245 for *PRF* method. Thus Table 7.2 show that the use of Word Embeddings combined to frequent closed patterns makes query expansion better than only considering patterns.

Pattern mining utilized for short query expansion was effective for most queries. Due to the shortness of queries and content of tweets, some queries give a poor performance in evaluations. That's why we used Word Embeddings to extend patterns and enrich the query. This additional method added to the query expansion model leads to an improvement compared to the pattern and baseline runs as shown in Fig. 7.2. Compared to the *baseline*, we have obtained significant improvements over the four measures : +191,30%, +218,01%, +134,43% and +168,5% respectively on P@10, P@30, MAP and nDCG@10.

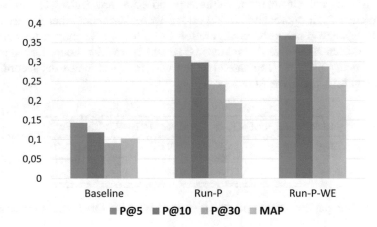

Fig. 7.2: Comparison of our runs with respect to the baseline run

Table 7.3 shows Precision at 30 (P@30), Mean Average Precision (MAP) and Reciprocal Precision (R-Prec) for our proposed approach and for the different expansion methods proposed by authors in [19]. The focus of the comparison is on this approach as since the authors used the concept of co-occurrence (by using frequent patterns) and have evaluated the effectiveness of the proposed approach on the same dataset (TREC 2011). *RUN1* denotes a run using term frequency without weighting. *RUN2* is a second run where the terms weighting is based on the occurrence of a term in the patterns. *RUN3* is the third run which is based on patterns only where the weighting of a pattern is performed using the weights of terms contained in the pattern. Finally, *RUN4* represents the combination of the second and the third run [19].

Metric/RunID	Run1A	Run2A	Run3A	Run4A	Run-P-WE
P@30	0.1347	0.1694	0.2034	0.1973	0.2878
MAP	0.0753	0.0486	0.0673	0.0486	0.2403
R-PREC	0.1114	0.0846	0.1191	0.1040	0.2475

Table 7.3: Comparison of our proposed method with the QE method proposed by [19]

Table 7.4 shows MAP, Normalized discounted cumulative gain (nDCG) and F-measure for the proposed approach compared to the query expansion method introduced by authors in [37]. This query expansion approach is based on multiple sources of external information. Authors proposed a new framework to expand users queries. Their approach is based on the *Word2Vec* model that we used. The run *NMF+Query* denotes the configuration using a negative matrix factorization (NMF). The run *NMF+W2V* combines NMF and *Word2Vec* model for expanding the initial query [37]. We compared our run with their different runs according to the MAP, nDCG, and F-measure scores.

Metric/RunID	NMF+Query	Word2Vec	NMF+W2V	Run-P-WE
MAP	0.036	0.093	0.027	0.2403
nDCG	0.226	0.219	0.272	0.4476
F-measure	0.101	0.039	0.092	0.2646

Table 7.4: Comparison of our proposed method with the QE method proposed by [37]

We observe that our proposed model gives a good performance compared to their runs. We have achieved some major improvements on the three measures: : +158,39%, +64,56% and +161,98% respectively on MAP, nDCG and F-measure.

Furthermore, different expansion approaches were the subject of participation in the TREC Microblog track 2011, we can mention the work proposed by authors in [21] who obtained better performance. This approach has yielded a precision of 0.3973 considering only 30 first results returned by the IR system, and a MAP score of 0.3157. It is based on a calculation of scores by modeling the relevant tweets by a Gaussian distribution and the irrelevant tweets by an exponential distribution. The authors also used the local expansion method *PRF*. The retrieval results are obtained using a language model as a IR model.

In overall, the experimental results show that the approach leveraging patterns and Word Embeddings outperforms the baseline and some methods of the literature. The formal concept lattice we used to extend queries selects the related terms which appear together in documents (tweets). Also, the extraction of interesting frequent patterns allows us to select the most important terms related to the initial query by considering the top N retrieved documents.

In our empirical study, we fixed the number of retrieved documents in the initial query to 500 (N). When this number increases to 1000, there is no significant improvement. We also have varied the minsup value for extracting patterns and have chosen 10 (i.e. 2%). We fixed $k = 3$ representing the number of most similar terms found by *cosine* similarity measure. In regards to *BM25* parameters, we have used the default configuration—i.e., $b = 0.75$ and $k1 = 1.2$. All experiments we report on evaluation metrics are performed using TREC EVAL.

7.6 Conclusion

In this chapter, we proposed a query expansion approach to overcome the shortness of user queries and tweets and enhance the quality of information retrieval in microblogs. The shortness of the microblogs and the queries may impact the quality of search. Our proposed method is based on frequent closed patterns and formal concept analysis. The frequent closed patterns are combined with word embeddings for finding the words that are most similar to the original query. The use of frequent closed patterns has allowed us to identify terms to further enrich the initial query by taking into account the co-occurrences of words. This method of local expansion is based on the first documents retrieved by the IR system for the selection of expansion terms. We combined the patterns with word embeddings to find the most similar terms to the patterns. The use of word embedding has allowed us to enrich the query with terms contained in the retrieval corpus. These new terms are words that appear in similar contexts of the corpus over the original query terms.

We conducted experiments and evaluated our proposed approach on TREC 2011 tweets collection. We compared its performance to some state-of-the-art approaches and to PRF approach. Results showed that the proposed method considerably improves the accuracy, the nDCG and the Mean Average Precision (MAP). Experiments conducted on this collection showed that combining patterns with word embeddings improves the quality of the returned search results. Results revealed the effectiveness of the proposed approach and show the interest of combining patterns and word embedding to enhance microblog search.

Expanding short queries is one of the most important solutions to improve the quality of tweets search results. It essentially consists of enriching the initial query by adding more similar and more significant terms, while keeping the same meaning in context. It is possible to integrate our expansion approach into a tweet search system that allows the users to reformulate their initial query. This can enhance the relevance of the results.

In our future work, it will be interesting to investigate temporal information presented in tweets. We also propose to integrate the location information while searching within the tweets where the query can be composed of region-of-interest (ROI) and text. We will further investigate external resources to compare our proposed method with external query expansion approaches.

References

1. Aggarwal N, Buitelaar P (2012) Query expansion using Wikipedia and DBpedia. In: CLEF (Online Working Notes/Labs/Workshop)
2. Agrawal R, Imieliński T, Swami A (1993) Mining association rules between sets of items in large databases. SIGMOD Rec 22(2):207–216. https://doi.org/10.1145/170036.170072
3. ALMasri M, Berrut C, Chevallet JP (2013) Wikipedia based Semantic Query Enrichment. In: Proceedings of the Sixth International Workshop on Exploiting Semantic Annotations in Information Retrieval, ESAIR '13, pp 5–8
4. Almasri M, Berrut C, Chevallet J (2016) A Comparison of Deep Learning Based Query Expansion with Pseudo-Relevance Feedback and Mutual Information. In: Advances in Information Retrieval - 38th European Conference on IR Research, ECIR 2016, Padua, Italy, March 20–23, 2016. Proceedings, pp 709–715
5. Amati G (2003) Probability models for information retrieval based on divergence from randomness. PhD thesis, University of Glasgow
6. Bai J, Song D, Bruza P, Nie Jy, Cao G (2005) Query expansion using term relationships language models for information retrieval. International Conference on Information and Knowledge Management, Proceedings
7. Carpineto C, Romano G (2012) A survey of automatic query expansion in information retrieval. ACM Comput Surv 44(1)
8. Codocedo V, Napoli A (2015) Formal Concept Analysis and Information Retrieval – A Survey. In: International Conference in Formal Concept Analysis - ICFCA 2015, Springer, Nerja, Spain, vol 9113, pp 61–77
9. Codocedo V, Baixeries J, Kaytoue M, Napoli A (2016) Contributions to the Formalization of Order-like Dependencies using FCA. In: What can FCA do for Artificial Intelligence?, The Hague, Netherlands

10. Diaz F, Mitra B, Craswell N (2016) Query expansion with locally-trained word embeddings. CoRR abs/1605.07891
11. Dogra N, Mulhem P, Goeuriot L, Amini MR (2018) Corpus d'entraînement sur les plongements de mots pour la recherche de microblogs culturels. In: COnférence en Recherche d'Informations et Applications - CORIA 2018, Rennes, France
12. Ganter B, Wille R (1999) Formal concept analysis: mathematical foundations. Springer Science
13. Gong Z, Cheang CW, Hou U L (2006) Multi-term Web Query Expansion Using WordNet. In: Database and Expert Systems Applications, Springer Berlin Heidelberg, Berlin, Heidelberg, pp 379–388
14. Han J, Cheng H, Xin D, Yan X (2007) Frequent pattern mining: current status and future directions. Data Min Knowl Discov 15(1):55–86
15. Hu J, Deng W, Guo J (2006) Improving retrieval performance by global analysis. In: 18th International Conference on Pattern Recognition, vol 2, pp 703–706
16. Järvelin K, Kekäläinen J (2002) Cumulated gain-based evaluation of IR techniques. ACM Trans Inf Syst 20(4):422–446
17. Jones KS, Walker S, Robertson SE (2000) A probabilistic model of information retrieval: Development and comparative experiments. Inf Process Manage 36(6)
18. Kotov A, Zhai C (2012) Tapping into knowledge base for concept feedback: Leveraging ConceptNet to improve search results for difficult queries. In: Proceedings of the Fifth ACM International Conference on Web Search and Data Mining, ACM, New York, NY, USA, WSDM '12, pp 403–412
19. Lau C, Li Y, Tjondronegoro D (2011) Microblog retrieval using topical features and query expansion. Proceedings of The Twentieth Text REtrieval Conference
20. Li W, Jones GJF (2017) Comparative evaluation of query expansion methods for enhanced search on microblog data: DCU ADAPT @ SMERP 2017 workshop data challenge. In: Proceedings of the First International Workshop on Exploitation of Social Media for Emergency Relief and Preparedness co-located with European Conference on Information Retrieval, pp 61–72
21. Li Y, Dong X, Guan Y (2011) HIT_LTRC at TREC 2011 Microblog Track. In: Text REtrieval Conference (TREC) 2011
22. Macdonald C, Ounis I (2007) Expertise Drift and Query Expansion in Expert Search. In: Proceedings of the Sixteenth ACM Conference on Conference on Information and Knowledge Management, CIKM '07, pp 341–350
23. Massoudi K, Tsagkias M, de Rijke M, Weerkamp W (2011) Incorporating query expansion and quality indicators in searching microblog posts. In: Proceedings of the 33rd European Conference on Advances in Information Retrieval, Springer-Verlag, Berlin, Heidelberg, ECIR'11, pp 362–367
24. Mikolov T, Chen K, Corrado G, Dean J (2013) Efficient Estimation of Word Representations in Vector Space. In: Proceedings of the International Conference on Learning Representations, pp 1–12
25. Mittal N, Nayak R, Govil MC, Jain KC (2010) Dynamic query expansion for efficient information retrieval. In: 2010 International Conference on Web Information Systems and Mining, vol 1, pp 211–215
26. Ounis I, Amati G, Plachouras V, He B, Macdonald C, Johnson D (2005) Terrier information retrieval platform. In: Proceedings of the 27th European Conference on Advances in Information Retrieval Research, Springer-Verlag, Berlin, Heidelberg, pp 517–519
27. Ounis I, Macdonald C, Lin J, Soboroff I (2011) Overview of the trec-2011 microblog track. In: In Proceedings of TREC 2011
28. Pal D, Mitra M, Bhattacharya S (2015) Exploring query categorisation for query expansion: A study. CoRR abs/1509.05567. arXiv:1509.05567

29. Pasquier N, Bastide Y, Taouil R, Lakhal L (1999) Efficient mining of association rules using closed itemset lattices. Inf Syst 24(1):25–46
30. Robertson SE, Walker S (1994) Some Simple Effective Approximations to the 2-Poisson Model for Probabilistic Weighted Retrieval. In: Croft BW, van Rijsbergen CJ (eds) SIGIR '94, Springer London, London, pp 232–241
31. Rocchio JJ (1971) Relevance feedback in information retrieval. In: Salton G (ed) The Smart retrieval system - experiments in automatic document processing, Englewood Cliffs, NJ: Prentice-Hall, pp 313–323
32. Roy D, Paul D, Mitra M, Garain U (2016) Using word embeddings for automatic query expansion. CoRR abs/1606.07608. arXiv:1606.07608
33. Sanderson M (2010) Test collection based evaluation of information retrieval systems. Foundations and Trends®in Information Retrieval 4(4):247–375
34. Silva PRC, Dias SM, Brandão WC, Song MAJ, Zárate LE (2017) Formal concept analysis applied to professional social networks analysis. In: Proceedings of the 19th International Conference on Enterprise Information Systems, Volume 1, Porto, Portugal, April, 2017, pp 123–134
35. Spink A, Wolfram D, Jansen J, Saracevic T (2001) Searching the web: The public and their queries. Journal of the American Society for Information Science and Technology 52:226 – 234
36. Xu J, Croft WB (1996) Query Expansion Using Local and Global Document Analysis. In: Proceedings of the 19th Annual International ACM SIGIR Conference on Research and Development in Information Retrieval, SIGIR '96, pp 4–11
37. Yang Z, Li C, Fan K, Huang J (2017) Exploiting multi-sources query expansion in microblogging filtering. Neural Network World 27:59–76
38. Zaki MJ, Hsiao C (2005) Efficient algorithms for mining closed itemsets and their lattice structure. IEEE Trans Knowl Data Eng 17(4):462–478
39. Zhai C, Lafferty J (2001) Model-based feedback in the language modeling approach to information retrieval. In: Proceedings of the Tenth International Conference on Information and Knowledge Management, ACM, NY,USA, pp 403–410
40. Zingla MA, Chiraz L, Slimani Y (2016) Short query expansion for microblog retrieval. Knowledge-Based and Intelligent Information & Engineering Systems: Proceedings of the 20th International Conference KES-2016 96(C)
41. Zingla MA, Latiri C, Mulhem P, Berrut C, Slimani Y (2018) Hybrid query expansion model for text and microblog information retrieval. Inf Retr Journal 21(4):337–367

Chapter 8
Scalable Visual Analytics in FCA

Tim Pattison, Manuel Enciso, Ángel Mora, Pablo Cordero, Derek Weber, and Michael Broughton

8.1 Introduction

Formal Concept Analysis (FCA) [88] derives a multiple-inheritance type hierarchy from a formal context. A *formal context* is a bigraph, consisting of a set of *object* vertices, a set of *attribute* vertices, and edges specified by a binary relation between these two sets. The "types" derived by FCA correspond to maximal bicliques in this bigraph, and are known as *formal concepts*. A formal concept consists of a set of objects, called its *extent*, and a set of attributes, called its *intent*, which are fully interconnected and jointly maximal. The set of formal concepts, when partially ordered by set inclusion on their extents, forms a complete lattice. This lattice can be efficiently represented as a single-source, single-sink, labelled, directed acyclic graph (DAG)—henceforth called the *lattice digraph*—whose vertices are formal concepts, and whose adjacency relation is the transitive reduction [1] of the ordering relation.

The resultant multiple-inheritance hierarchy of formal concepts constitutes a useful generalisation of a hierarchy for applications such as the storage and retrieval of data objects using keywords or tags, the partial ordering of closed frequent item sets in association mining, or the representation of a Description Logic subsumption hierarchy [75]. Accordingly, FCA has been widely applied in such disparate fields as information retrieval, knowledge discovery and knowledge representation [66, 68, 91].

Formal Concept Analysis is an analytic technique suitable for organisations at different levels of maturity in information exploitation (see e.g. [35, 76]). For those who primarily retrieve and read unstructured text, the bi-adjacency matrix of the context bigraph is closely related to the "bag of words" representation common to a number

T. Pattison (✉) • D. Weber • M. Broughton
Defence Science and Technology Group, Adelaide, SA, Australia
e-mail: tim.pattison@dst.defence.gov.au

M. Enciso • Á. Mora • P. Cordero
Universidad de Málaga, Málaga, Spain

© The Author(s), under exclusive license to Springer Nature Switzerland AG 2022
R. Missaoui et al. (eds.), *Complex Data Analytics with Formal Concept Analysis*,
https://doi.org/10.1007/978-3-030-93278-7_8

167

of statistical techniques for natural language processing, such as Latent Semantic Analysis [46]. For those analysing tagged data, including various forms of social media, the tags are attributes associated with the media objects of interest. FCA constitutes a form of association mining for structured data such as the membership of people in organisations or communities of interest, producing a set of implications amongst the chosen attributes. For organisations aspiring to formal knowledge representation for automated reasoning, the output of FCA is an empirically-derived subsumption hierarchy whose incorporation into Description Logics has attracted considerable research interest [75].

8.1.1 Scalable Visual Analytics in FCA

"Visual analytics is the science of analytical reasoning facilitated by interactive visual interfaces," which, inter alia, "seeks to marry techniques from information visualisation with techniques from computational transformation and analysis of data" [85]. We adopt a visual analytic approach to FCA by combining computational analysis with interactive visualisation. Scalability is a key challenge for visual analytics. Algorithms must scale to large data sets, visualisations must make efficient and intelligible use of screen real-estate, and both must be responsive for interactive use. The number of formal concepts and the number of implications derived from a formal context are both bounded above by exponential functions of the number of attributes in that context. Moreover, the set of all valid implications in a formal context is highly redundant. The latter problem has motivated essential research into an equivalent basis set of implications, having reduced cardinality, from which any valid implication can be derived. The first such basis was proposed by Duquenne and Guigues [36].

Consequently, four fundamental challenges confront those who wish to scale FCA to the interactive analysis of large data sets: the time required to enumerate the vertices, arcs and labels of the lattice digraph; the difficulty of meaningful and responsive user interaction with a large lattice digraph; the time required to enumerate (a basis for) all valid implications; and the difficulty of meaningful and responsive user interaction with a large set of implications to discover those which deliver insight.

This chapter focuses on scaling interactive FCA to handle the characteristic volume of big data, rather than its velocity, variety or veracity. We assume that the number $|G|$ of objects is larger than the number $|M|$ of attributes, in which case the number of formal concepts is bounded above by $2^{|M|}$. We further assume that while $|G|$ may be very large, either $|M|$ remains moderate or the formal context is relatively sparse, so that the threatened exponential explosion in the number of formal concepts is not realised. Godin et al. [34] provide empirical support for this assumption in the domain of information retrieval. Aggregating all objects into a single formal context is justified provided all objects are sampled from the same unknown but stationary joint probability distribution (see e.g. [57]) over the attributes. If there is reason to believe that this distribution is instead slowly time-varying, periodically discarding older objects and rebuilding the concept lattice may be more appropriate.

8.1.2 Organisation

This chapter is organised as follows. Section 8.2 provides a graph-theoretic introduction to Formal Concept Analysis. Section 8.3 introduces the topic of visual analytics and its application to Formal Concept Analysis. Section 8.4 undertakes a brief survey of techniques aimed at improving the scalability of FCA for more responsive visualisation and interaction. Section 8.5 briefly presents three techniques and associated software prototypes through which the Defence Science and Technology (DST) Group has addressed selected aspects of FCA scalability. Section 8.6 describes the use of data visualisation by the Universidad de Málaga to help users find meaningful implications. The chapter concludes with a summary along with proposals for future research and engineering challenges.

8.2 Graph-Theoretic Introduction to FCA

Whereas the foundations of Formal Concept Analysis are typically expounded in mathematical terms (see e.g. [32]), we choose a graph-theoretic exposition which aligns more naturally with our visualisation objectives.

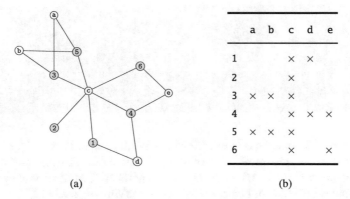

	a	b	c	d	e
1			×	×	
2			×		
3	×	×	×		
4			×	×	×
5	×	×	×		
6			×		×

(a)   (b)

Fig. 8.1: Two representations of a formal context. (**a**) Drawing of context bigraph. (**b**) Bi-adjacency table

8.2.1 Formal Context

A formal context $\mathbb{K} = (G, M, I)$ is a bipartite graph, or *bigraph*, with object vertex set G, attribute vertex set M, and undirected edge set $I \subseteq G \times M$. Each object and attribute vertex has a unique label which derives from the domain of application. For an information retrieval domain, for example, the object labels may be document titles and the attribute labels keywords. The context bigraph can be visualised using standard graph drawing techniques (Fig. 8.1a), or its bi-adjacency matrix tabulated (Fig. 8.1b). While the latter approach predominates amongst FCA practitioners, Pattison and Ceglar [61] described a layout of the context bigraph which clusters the constituent objects and attributes of a formal concept.

A sub-context $\underline{\mathbb{K}} = (\underline{G}, \underline{M}, \underline{I})$ of the formal context (G, M, I) is a bigraph consisting of a subset $\underline{G} \subseteq G$ of its objects, a subset $\underline{M} \subseteq M$ of its attributes, and the subset $\underline{I} = I \cap (\underline{G} \times \underline{M})$ of its edges adjacent to those object and attribute vertices.

8.2.2 Formal Concepts

A formal concept $(\mathscr{E}, \mathscr{I})$ of the context $\mathbb{K}$ consists of a subset $\mathscr{E} \subseteq G$ of objects, called the *extent*, and a subset $\mathscr{I} \subseteq M$ of attributes, called the *intent*, which form a maximal biclique [33]. A *biclique* $(\mathscr{E}, \mathscr{I})$ is a fully bi-connected subgraph: $\mathscr{E} \times \mathscr{I} \subseteq I$. It is *maximal* if no proper superset $\overline{\mathscr{E}} \supset \mathscr{E}$ satisfies $\overline{\mathscr{E}} \times \mathscr{I} \subseteq I$ and no proper superset $\overline{\mathscr{I}} \supset \mathscr{I}$ satisfies $\mathscr{E} \times \overline{\mathscr{I}} \subseteq I$. A formal concept need not be a *proper* biclique [33], since it may have empty intent or extent.

Example 8.1 In the example context of Fig. 8.1a, $(\mathscr{E}, \mathscr{I}) = (\{3, 5\}, \{a, b, c\})$ is a formal concept since all attributes in $\mathscr{I} = \{a, b, c\}$ are connected to both objects in $\mathscr{E} = \{3, 5\}$, and neither set shares any additional neighbours. The attribute vertices $m \in \mathscr{I}$ and object vertices $g \in \mathscr{E}$ are highlighted in Fig. 8.2a with magenta and orange fill, respectively, on a pink background. The maximal but improper biclique $(\varnothing, M)$ is a formal concept in this context.

Definition 8.1 The *intent operator* $' : \mathscr{P}(G) \to \mathscr{P}(M)$ maps any set $\mathscr{A} \subseteq G$ of object vertices to the maximal set $\mathscr{A}' \subseteq M$ of attribute neighbours satisfying $\mathscr{A} \times \mathscr{A}' \subseteq I$. The *extent operator* $' : \mathscr{P}(M) \to \mathscr{P}(G)$ maps any set $\mathscr{B} \subseteq M$ of attribute vertices to the maximal set $\mathscr{B}' \subseteq G$ of object neighbours satisfying $\mathscr{B}' \times \mathscr{B} \subseteq I$.

Here $\mathscr{P}(\mathscr{S})$ denotes the powerset of a finite set $\mathscr{S}$. Since it is obvious from the context which of these two operators is intended, the same symbol $'$ usually suffices for both. The intent $\mathscr{I} = \mathscr{E}'$ and extent $\mathscr{E} = \mathscr{I}'$ of a concept are closed under the composition $''$ of these two operators, since $\mathscr{I} = \mathscr{E}' = \mathscr{I}''$ and $\mathscr{E} = \mathscr{I}' = \mathscr{E}''$. Formal Concept Analysis involves, at a minimum, enumerating all such closed sets of attributes or objects for a formal context.

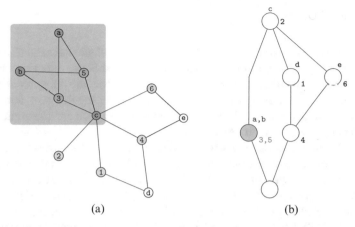

(a) (b)

Fig. 8.2: Visualisation of a formal concept. (**a**) Context bigraph. (**b**) Lattice digraph

8.2.3 Concept Lattice Digraph

FCA produces a set $\mathfrak{B}$ of formal concepts which are partially-ordered, according to convention, by the relation $\leq$ denoting extent set inclusion. Poset $(\mathfrak{B}, \leq)$ forms a complete lattice [22], so that any subset of concepts has a unique least upper bound or *supremum*, and a unique greatest lower bound or *infimum*. The lattice has supremum (G, G') and infimum (M', M), respectively. For the example formal context of Fig. 8.2a, these are $(\{1, 2, 3, 4, 5, 6\}, \{c\})$ and $(\varnothing, \{a, b, c, d, e\})$.

$(\mathfrak{B}, \leq)$ can be efficiently represented as a DAG, whose vertices are formal concepts, and whose directed adjacency relation is the cover relation [22]. The *cover* relation is the transitive reduction [1] of the ordering relation $<$, and is directed from the lesser to the greater concept. Two distinct concepts are comparable iff there is a directed path between their corresponding vertices in the lattice digraph. The lattice digraph has a single source vertex, representing the infimum of the lattice, and a single sink vertex, representing the supremum.

The *join* or *least upper bound* on a set $\mathscr{S} \subseteq \mathfrak{B}$ of concepts is the concept at which directed paths from all concepts in $\mathscr{S}$ first converge. The intent of this concept is the intersection of the intents of the concepts in $\mathscr{S}$. Similarly, the *meet* or *greatest lower bound* on $\mathscr{S}$, denoted $\bigwedge_{x \in \mathscr{S}} x$, is the concept at which directed paths to all concepts $x \in \mathscr{S}$ first diverge. The extent of $y = \bigwedge_{x \in \mathscr{S}} x$ is the intersection of the extents of the concepts in $\mathscr{S}$.

The poset consisting of the elements of $\mathfrak{B}$ ordered instead by intent set inclusion is dually isomorphic to $(\mathfrak{B}, \leq)$: the arcs of its corresponding digraph are simply reversed. This *Galois connection* [80] between the two posets provides a formal foundation for Formal Concept Analysis.

8.2.4 Line Diagram

A line diagram [31] is a drawing of the lattice digraph in which the vertical component of each arc is upwards on the page. This convention aids the interpretation of the partial ordering and obviates the need to explicitly indicate the direction of each arc. The source [sink][1] vertex appears at the bottom [top] of the diagram, and all other vertices are assigned intermediate vertical coordinates which are constrained by the direction convention for their adjacent arcs.

Each attribute is associated with a unique concept whose extent is the set of objects adjacent to that attribute in the context bigraph. Similarly, each object is associated with a unique concept whose intent is the set of adjacent attributes. These are known as *attribute* and *object concepts*, respectively. Each attribute [object] concept is labeled in the line diagram with the corresponding attributes [objects]. Attribute [object] labels are placed above [below] the vertical centreline of the labelled concept. A concept inherits the attributes [objects] appearing as labels on comparable concepts above [below] it in the line diagram.

Example 8.2   Figure 8.2b shows the line diagram resulting from FCA of the context of Fig. 8.2a. The top vertex in Fig. 8.2b is the attribute concept for attribute c and the object concept for object 2. The vertex having attribute label set $\{d\}$ and object label set $\{1\}$ corresponds to the concept having extent $\{1,4\}$ and intent $\{c,d\}$. In addition to its own object and attribute labels, it inherits attribute c from its upper neighbour and object 4 from its lower neighbour.

The line diagram encapsulates the concepts, order relation, Galois connection and, as we will see in Sect. 8.2.7, implications of a formal context in one simple representation. Accordingly, Eklund and Villerd [26] identified the line diagram as a key enabler for various methods in conceptual knowledge processing. Novices were found capable of reading small line diagrams to accomplish simple information retrieval tasks [27]. However, properly interpreting all of the available information requires a good working knowledge of FCA, and becomes increasingly difficult with increasing digraph size and complexity. Since the underlying digraph model cannot be simplified without discarding information, improvements to the aesthetic qualities and ultimate utility of the line diagram are the exclusive preserve of the chosen graph drawing and interaction technique(s).

8.2.5 Simplifying Implications

The knowledge encapsulated in the line diagram can also be represented as a set of attribute implications. An attribute implication is an expression $\mathscr{L} \rightarrow \mathscr{R}$ where

---

[1] Square brackets are used throughout this chapter to indicate that a sentence is true both when read without the bracketed terms and when read with each bracketed term substituted for the term which precedes it.

$\mathscr{L}, \mathscr{R} \subseteq M$ and we say that $\mathscr{L} \rightarrow \mathscr{R}$ holds in $\mathbb{K}$, or that $\mathbb{K}$ is a model for $\mathscr{L} \rightarrow \mathscr{R}$, whenever $\mathscr{L}' \subseteq \mathscr{R}'$. The implication $\mathscr{L} \rightarrow \mathscr{R}$ is composed of two parts: its antecedent or left-hand side (LHS) $\mathscr{L}$ and its consequent or right-hand side (RHS) $\mathscr{R}$. In contrast with the lattice digraph, a set of implications can be redundant and may be simplified through symbolic manipulation.

For completeness, we seek a set of attribute implications from which all other valid implications over the formal context $\mathbb{K}$ can be derived. Efficient algorithms exist for the production of such *implicational bases* [67]. To reduce the volume of information presented to the user, we seek a basis which is both minimal—i.e. has the smallest number of implications—and optimal—i.e. has the smallest sum over all implications of LHS and RHS cardinalities [87]. Developed by our team at the University of Málaga, Simplification Logic [53] can be applied automatically to a given implicational basis to produce, with quadratic cost, a set of implications having closed right-hand sides and left-hand sides which are minimal generators of the right-hand sides [10, 21]. Such a basis is both minimal and optimal.

8.2.6 Visualising Implications

This subsection briefly introduces an interactive table view as an example visualisation of a set of implications, deferring until Sect. 8.6 a review of other available techniques and the exposition of a novel approach.

	a	b	c	d	e
1	×	×			
2		×		×	×
3		×	×	×	
4			×		×
5				×	×

Table 8.1: Example formal context

Serial	$\mathscr{L}$	$\rightarrow$	$\mathscr{R}$	$\mathscr{R} \setminus \mathscr{L}$	$|\mathscr{R}'|$
1	$\{a\}$	$\rightarrow$	$\{a, b\}$	$\{b\}$	1
2	$\{b, c\}$	$\rightarrow$	$\{b, c, d\}$	$\{d\}$	1
3	$\{c, d\}$	$\rightarrow$	$\{b, c, d\}$	$\{b\}$	1
4	$\{b, e\}$	$\rightarrow$	$\{b, d, e\}$	$\{d\}$	1
5	$\{a, b, d\}$	$\rightarrow$	$\{a, b, c, d, e\}$	$\{c, e\}$	0
6	$\{b, c, d, e\}$	$\rightarrow$	$\{a, b, c, d, e\}$	$\{a\}$	0

Table 8.2: Duquenne-Guigues canonical basis for implications on the context of Table 8.1

The implications of the formal context in Table 8.1 can be enumerated algorithmically and tabulated as per Table 8.2 for exploration by the user. An interactive table view constitutes a familiar interface through which the user can sort, filter, inspect and visually compare rows, which in this case correspond to implications. Implications can be sorted, for example, by the cardinality or lectic order of their antecedent, consequent or abbreviated consequent sets, or filtered by constraining set membership. Sorting and filtering operations can be useful for managing potentially large numbers of implications. A range of objective measures, including support and lift, has also been defined [50] for directing the user's attention to "interesting" implications; these interestingness scores can be added to the table, and used thereafter for sorting and filtering. Figure 8.12 shows a screenshot of an interactive table view of implications which supports sorting and filtering by selected interestingness measures. Computational assistance, such as highlighting set intersections or differences, is advisable when comparing attribute sets containing more than a handful of elements.

8.2.7 Coordinating Views of Implications and Concepts

Whilst several tools support the presentation of both concept lattices and sets of implications [2, 52, 90], they provide no mechanism for view coordination between them. By relating an implication to corresponding elements of the concept lattice, this subsection provides the potential basis for such view coordination.

Definition 8.2 An implication $\mathscr{L} \to \mathscr{R}$ is said to be **full** if $\mathscr{L} \subsetneq \mathscr{L}'' = \mathscr{R}$.

A full implication is manifested in a drawing of the lattice digraph as follows.

Definition 8.3 The **consequent concept** of a full implication $\mathscr{L} \to \mathscr{R}$ is the formal concept having intent $\mathscr{L}'' = \mathscr{R}$.

The digraph vertex corresponding to the consequent concept serves as a graphical anchor for, and could mediate user interaction with, a full implication. Unambiguous specification of an implication, however, requires stipulation of not only its consequent, but also its antecedent. Whereas no formal concept has intent $\mathscr{L}$, the constituent attributes $m \in \mathscr{L}$ of the antecedent are represented in the lattice digraph by their corresponding attribute concepts (m', m'').

Proposition 8.1 *Let $\mathscr{L} \to \mathscr{R}$ be a full implication of the formal context (G, M, I). Then its consequent concept is*

$$(\mathscr{R}', \mathscr{R}) = (\mathscr{L}', \mathscr{R}) = \bigwedge_{m \in \mathscr{L}} (m', m'')$$

The consequent concept of a full implication is the meet of the attribute concepts for its antecedent attributes $m \in \mathscr{L} \subseteq \mathscr{R}$. A full implication can therefore be graphically validated by confirming that: downward paths in the line diagram from the antecedent attribute concepts first converge at the consequent concept; and an upward path exists from the consequent concept to each attribute concept of the abbreviated consequent.

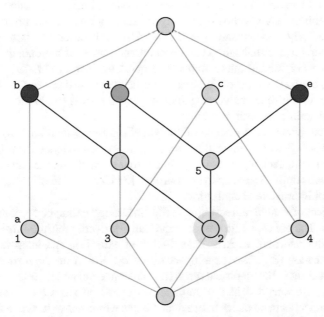

Fig. 8.3: Lattice with attribute concepts for antecedent and abbreviated consequent attributes of full implication $\{b, e\} \to \{b, e, d\}$ filled red and green, respectively

The attribute concept for each attribute $m \in \mathscr{R}$ can be coloured in the line diagram according to whether the attribute belongs to the antecedent $\mathscr{L}$ or the abbreviated consequent $\mathscr{R} \setminus \mathscr{L}$. Following international maritime and aeronautical conventions for left and right, we colour these red and green respectively. Figure 8.3 illustrates this colour convention for the implication $\{b, e\} \to \{b, e, d\}$ of the example formal context in Table 8.1. The concept whose intent is the consequent is shown with yellow halo. Convergence at the haloed vertex of downward paths from the two red vertices confirms that the consequent is the meet of the attribute concepts for members of the antecedent. An upward path from the haloed vertex to the green vertex confirms that attribute d is implied by the presence of the antecedent attributes b, e.

8.3 Introduction to Visual Analytics

In their review of Big Data applications, techniques and technologies, Chen and Zhang [17] identify data analysis and data visualisation, inter alia, as significant challenges. The high volume of big data necessitates the use of automated algorithmic analysis wherever analytic objectives are anticipated and well defined. Its variety, on the other hand, makes ad hoc analysis inevitable, to which interactive information visualisation is better suited. Sucharitha et al. [82] found the principal benefits of information visualisation for the analysis of big data to include: improved decision-making; better ad-hoc data analysis; and improved collaboration/information sharing. While visualisation is traditionally considered the last step of data analysis, interactive visualisation can combine the power of big data analytics with the advantages of human interaction in the quest for insight [17, 79]. The variety of big data, coupled with evolving analytic objectives, further requires the flexible combination of multiple analytic techniques, whose results can be interfaced to the user through multiple coordinated views.

"Visual analytics is the science of analytical reasoning facilitated by interactive visual interfaces," which, inter alia, "seeks to marry techniques from information visualisation with techniques from computational transformation and analysis of data" [85]. We adopt a visual analytic approach to FCA by combining computational analysis with interactive visualisation.

The subject of analytical reasoning is the "meaning" extracted from the available information. The concept lattice is an empirically derived, multiple-inheritance type hierarchy which formalises Aristotle's duality of extension and intension [77] as a Galois connection [80]. The meaning extracted by FCA from the formal context is therefore inherent in the empirical types (formal concepts) and their partial ordering. Meaning is also encoded in the implications extracted from the formal context. These are logical dependencies between attributes which are reflected in the absence from the formal context of some attribute combinations. Such dependencies map the elements of the Boolean lattice of attributes onto fewer empirical types.

8.3.1 Algorithmic Analysis

Visual analytics involves computational transformation beyond simple querying and filtering of the available information, although even these can prove challenging for interactive analysis of big data [17]. Algorithms exist for enumerating the concepts of a formal context and their partial ordering, as well as the implications amongst attributes. The interested reader is referred to reviews of FCA algorithms by Kuznetsov and Obiedkov [44] and Priss [68] and of implication algorithms by Poelmans et al. [67]. The time required to enumerate all formal concepts of a formal context is bounded above by a polynomial function of the number of formal concepts, which is in turn bounded above by an exponential function of the number of objects and attributes. The time required to calculate the transitive reduction of the ordering relation is also polynomial in the number of concepts and the cardinality of the relation. A number of FCA algorithms enumerate concepts and calculate the transitive reduction simultaneously [18, 49].

One approach to tackling the inherent computational complexity of FCA is to partition the context, perform FCA on each resultant sub-context, and then combine the results. Known as "divide and conquer", this approach allows FCA to be mapped onto multiple independent processors, with each processor performing FCA on a sub-context which is significantly smaller than the overall context. Several FCA algorithms use this approach for the enumeration of concepts [7, 11, 65].

8.3.2 Graph Drawing

Graph drawing is a form of visual analytics in that it marries interactive visualisation with the computational transformation and analysis of graph data. Visualisation of the lattice digraph as an interactive line diagram (see e.g. [2, 43]) necessarily constitutes visual analytics, since FCA is required to compute the lattice digraph from the context bigraph. Visualisation of the context bigraph and lattice digraph can also involve computational techniques such as force-directed layout of the former, or layered drawing of the latter so as to minimise edge crossings [26, 43, 61, 91]. The computation required typically scales poorly with the number of graph vertices and edges. Portraying the computed properties of a graph or of its elements, such as biclique membership or vertex centrality measures in the context bigraph [61], may similarly be considered to involve visual analytics.

8.3.3 Information Visualisation

Many types of information, such as medical tomography and weather data, have inherent two- or three-dimensional geometry arising from the domain of application. Whilst this geometry has the benefit that it is intuitive to users, it also heavily constrains how the information is visualised. Abstract information such as graphs, on the other hand, can be computationally mapped into one or more geometries designed to best support user exploration and decision-making, albeit at the expense of additional training which is required to develop user intuition in such abstract spaces.

For quantitative or ordinal data, assigning a geometry may simply involve choosing two or three aspects of the data and treating them as the dimensions of a Cartesian space. In Sect. 8.6.2, the "data" being visualised are instead derived from a set of implications, which are in turn computed from a formal context. The two spatial dimensions reflect quantitative properties of implications, while two other properties are mapped to glyph size and colour. The computational transformation of the formal context into implications for the purpose of visualisation once again justifies the label of "visual analytics".

8.3.4 Multiple Coordinated Views

Multiple views of the same data are often required to support decision-making or exploratory analysis. For example, a decision may require simultaneous consideration of the results of multiple analyses, or the data being explored may be inherently high-dimensional. To reduce the cognitive load on the user and facilitate insight, those multiple views should be coordinated. For example, filters applied to the data through interaction with one of the views are applied to all views, and data elements selected in one view are selected in all views in which they appear. The use of multiple coordinated views is accordingly well established in information visualisation (see e.g. [14, 63, 72]).

Users who are unfamiliar with FCA need to understand, explain and trust the transformation from the formal context to the concept lattice, and thereby develop their intuition regarding the concept lattice and its meaning. Coordinated views of the context bigraph and lattice digraph are therefore provided in the CARVE prototype [61, 64, 65]—see Sect. 8.5.1—albeit with coordination limited to using the same sub-context for both views. Additional forms of coordination are of course possible. For example, selection of attribute [object] vertices in the context bigraph might be coordinated with the selection of the corresponding attribute [object] concepts in the digraph, while selection of their neighbourhood might select the corresponding downset [upset].

In Sect. 8.2.7, inspection of an individual implication was facilitated through its representation in a drawing of the lattice digraph as the consequent concept and the attribute concepts for its constituent attributes. The potential utility of a table view for filtering and sorting a set of implications was also described. In Sect. 8.6, complementary data visualisations will be described which are designed to aid the exploratory user in identifying insightful implications. These multiple views should be coordinated, so that for example selection of a row of the table highlights the corresponding elements of the lattice digraph.

8.3.5 Tight Coupling

In its simplest form, visual analytics involves algorithmic analysis of a data set followed by visualisation of the results. For large formal contexts, however, the delay in generating the entire lattice digraph may relegate user interaction to the visualisation phase. Alternatively, as each new concept is generated, it can be added to the digraph representing the partial order amongst the concepts generated to date, allowing the user to commence familiarisation with the output. This may provide sufficient information for the user to steer the FCA algorithm to prioritise the generation of concepts of interest, or in fact to terminate the process of concept enumeration once their analytical objectives have been achieved. Such tight coupling between the FCA algorithm and visualisation of the lattice digraph is explored in Sect. 8.5.2.

8.4 Layout, Visualisation and Interaction

We saw in Sect. 8.1 that scalability is a key challenge for visual analytics. This section provides a brief survey of techniques aimed at improving the scalability of visual analytic FCA, and thereby affording the user more responsive and intelligible visualisation and interaction.

8.4.1 Reducing Digraph Size

The most obvious approach to improving line diagram layout, visualisation and interaction, is to reduce the number of formal concepts [67]. Any proper sub-lattice has fewer concepts than the full concept lattice, although prior to construction of the latter, options for nominating the supremum and/or infimum of the sub-lattice are constrained. The Attribute Object Concept (AOC) poset consists of only the attribute and object concepts, whose number scales linearly with the size of the context. The AOC poset can be visualised as a labelled line diagram, although care is required in the interpretation of meets and joins, since this poset is not a lattice. Another means of reducing the number of concepts is to impose a threshold on extent set cardinality, so that screen real-estate and user attention are reserved for formal concepts which represent suitably large subsets of the objects in the formal context, and which are typically higher in the multiple-inheritance type hierarchy. The partial order amongst these frequent closed item sets is referred to as an *iceberg* lattice. Algorithms exist [81] which exploit the monotonicity of the constraint on extent cardinality to expedite enumeration of the formal concepts.

8.4.2 Layout of Line Diagram

Rival [71] surveyed theoretical results and research directions regarding the properties of upward line drawings of partially-ordered sets. Ganter and Wille [32] cite several earlier articles on automated lattice drawing, and describe the automated drawing of additive line diagrams. An (object-)additive line drawing assigns to each supremum-irreducible element a vector having a positive vertical component, and to each concept a position given by the sum of the vectors corresponding to the supremum-irreducible objects in its extent (see e.g. [43, 91]). Techniques have been proposed [30, 92] to mitigate the problem of vertices overlapping with each other, or with edges to which they are not incident. Di Battista et al. [23] describe both force-directed methods for laying out general graphs and techniques for the layered drawing of directed graphs. Force-directed methods have also been used to lay out the vertices of a lattice DAG in three dimensions [29, 40].

Standard algorithms [23, 83] and genetic variants (see e.g. [56]) exist for assigning the vertices of a DAG to layers and ordering them within each layer to improve aesthetic criteria such as edge crossings. For a lattice digraph, layer assignment is constrained by the maximum path length of a vertex from the source, and to the sink, vertex. The assignment of vertices to layers in the line diagram is typically under-constrained by the partial order, so that both layer assignment and horizontal order within a layer can be permuted when adjusting the graph layout to optimise aesthetic criteria. This graph layout problem has combinatorial complexity [23]. However, preliminary results [61] suggest that reasonable edge crossing reduction can be achieved with polynomial complexity using multi-dimensional scaling of resistance distance in the context bigraph. The result is a form of seriation[2] [48] of the lattice *atoms* and *coatoms*—respectively upper neighbours of the infimum and lower neighbours of the supremum—from which the horizontal positions of all remaining concepts can be derived [60, 61].

8.4.3 Interactive Visualisation

Usability testing of FCA applied to the management of email has demonstrated that users can successfully interpret small line diagrams [27]. However, the potential combinatorial explosion of concepts with increasing size of the formal context poses challenges for the layout and visualisation of, as well as interaction with, the lattice digraph. On-demand construction and layout of the entire lattice digraph cannot be achieved in interactive timescales for large lattices, so either prior or user-guided construction and layout is required to support responsive interaction.[3] For contexts of even moderate size, the potentially large number of resultant vertices and arcs compete for limited screen real estate and challenge user comprehension.

To help the user manage this problem of scale, interactive exploration, as opposed to static presentation, of the line diagram is essential. Information visualisation techniques such as filtering, pan and zoom, focus-plus-context, details-on-demand and structural navigation [15] can support user interaction with, and comprehension of, large graphs [41]. For example, geometric zooming or distortion of the line diagram [16] can help allocate more screen real-estate to an area of interest. Alternatively, structural navigation of the lattice digraph can be facilitated by presentation of the immediate graph neighbourhood of the current vertex [16, 24, 89], possibly combined with an overview showing where that vertex resides in the full lattice. A third option is an interactive version of nested line diagrams [16, 88], in which each vertex serves as a container within which to display the line diagram for the same object set and (some of) the remaining attributes. In many applications, however, it is not clear *a priori* how best to group the attributes, or how to order the groups for nesting.

---

[2] We are indebted to Professor B. Ganter for drawing our attention to the terminology and literature of seriation.

[3] User-guided construction of the lattice digraph is addressed in Sect. 8.5.2.

8.4.4 Discovering or Imposing Tree Structure

A range of mature visualisation and interaction techniques exist for tree, as opposed to lattice, data structures [45, 69, 78]. Operating system interfaces for the structural navigation of directory hierarchies are ubiquitous, and user intuition is accordingly well established [16]. This intuition can be exploited for visualisation of the concept lattice digraph, provided that a tree structure can be discovered in, or imposed on, the graph. Any spanning tree of the lattice digraph, which is rooted at the source or sink vertex, would arguably serve this purpose. Melo et al. [51] investigate various criteria by which a single parent can be chosen for each concept. Rather than pruning the lattice digraph, Andrews and Hirsch [4] base their *concept tree* on the order in which concepts are discovered by the In-Close2 algorithm for concept enumeration.

A more general approach to imposing tree structure on a graph to facilitate user interaction is hierarchical clustering or partitioning of its vertex set [41]. Hierarchical clustering involves the recursive application of a graph clustering algorithm to the clusters (sub-graphs) it identifies. Graph clustering involves optimising some measure of cluster quality, such as modularity, which takes into account factors such as the number, or total weight, of intra- versus inter-cluster links. Whilst the global optimisation of modularity is NP complete, sub-optimal solutions can be computed for large graphs in responsive timeframes [19]. A range of techniques and tools exist for browsing hierarchically clustered graphs. Amongst these are structural zooming on inclusion layouts [69], and the GrouseFlocks environment [5] which supports the use and modification of multiple hierarchical clusterings on the same graph.

Despite its demonstrated scalability to large lattices [4], presenting the concept lattice as a concept tree rooted at the supremum has several disadvantages. Most significantly, it conceals from the user the multiple inheritance of attributes. It also proliferates object and attribute labels, and requires the dynamic relocation of object labels whenever a branch is interactively expanded or collapsed [4]. Whereas any given vertex will typically lie on multiple directed paths from the source [to the sink] of the lattice digraph, the corresponding path in a spanning tree is unique and arbitrary. To make it easier to purposefully locate a vertex of interest, or more likely that such a vertex might be encountered during less goal-directed user exploration, each vertex, along with the sub-lattice (tree) of which it is the supremum, can be replicated on demand under each of its parent vertices [16, 54]. By replicating entire sub-lattice concept trees, however, this approach: compromises the scalability reported by Andrews and Hirsch [4]; requires either redundant storage or behaviour not available in commodity tree browsers; and necessitates visual or interactive cues to indicate the identity of replicated concepts.

8.4.5 Demand for Enhanced Tool Support

Tags are now widely used on social media platforms, inviting better interfaces for organising social media content. Similarly, there is a clear trend in operating system interfaces towards tagging and querying rather than navigation of a hierarchical file system. Users typically require assistance in recalling or constructing a set of tags with which to retrieve a suitably small set of objects which contains the object(s) of interest. This trend will drive a demand for well-designed user interfaces through which, like trees before them, multiple-inheritance hierarchies become intuitive with use. The conceptually simple generalisation of a tree to allow a vertex to have multiple parents poses significant challenges for user navigation. More generally, scalable visualisation and interaction of multiple-inheritance hierarchies, and in particular of the digraph produced by FCA, remains an open challenge.

8.4.6 Implications

Any valid implication can be derived using an axiomatic system, such as Armstrong's rules [4] or Simplification Logic [20], from those in a suitable basis set [73, 74], whose cardinality is typically significantly smaller than the set of all valid implications. Enumeration, validation and exploration of only the implications in such a basis will therefore scale to somewhat larger contexts. As discussed in Sect. 8.2.7, a table view can be used to sort and filter these basis implications. Various interestingness measures [50] can be added to the table for use in sorting and filtering to direct user attention to potentially interesting implications. The data visualisations to be described in Sect. 8.6.2 can also be coordinated with this table view to assist the user in identifying implications of interest. The use of minimal generators as the left-hand sides, and abbreviated consequents as the right-hand sides, of implications ensures that implications contain no more attributes than necessary, thereby conserving screen real-estate.

8.5 Three FCA Prototypes

This section briefly presents three software prototypes which the DST Group has developed to address aspects of this scalability challenge.

8.5.1 Hierarchical Parallel Decomposition

CARVE [61, 64, 65] supports interactive analysis of large formal contexts by discovering and exploiting hierarchical structure, which we have previously identified in bibliographic contexts, and which Bhatti et al. [13] found in software systems. That hierarchical structure is used to expedite and enhance both the layout of, and user interaction with, the concept lattice. The CARVE algorithm [65] discovers a hierarchical decomposition of amenable contexts and of the corresponding lattice digraph.

It produces a tree, representing both a partial parallel decomposition of the lattice digraph [23] and a corresponding decomposition of the context bigraph, along with the digraph itself. The decomposition tree for the example context in Fig. 8.4a is shown in Fig. 8.4b. CARVE uses this tree both to divide and conquer the computational problem of laying out the digraph as a line diagram, and as a coordinated view to facilitate user interaction with the context bigraph and lattice digraph.

Each vertex of the tree returned by the CARVE algorithm corresponds to a sub-context identified during hierarchical decomposition of the formal context, and to the lattice digraph for that sub-context. This tree can be drawn using an inclusion layout, in which each vertex of the tree is represented as a container within which the containers representing descendant tree vertices are nested. In Figs. 8.4a and 8.4c, these nested containers are shown as coloured boxes whose colour is that of the corresponding vertex in the decomposition tree in Fig. 8.4b. Each leaf vertex of this tree serves in Fig. 8.4c as a container for the line diagram of the corresponding trivial or otherwise indivisible sub-lattice digraph. The use of these containers to enclose the corresponding bigraph and digraph vertices gives rise to the acronym CARVE, which stands for Context Analysis through Recursive Vertex Enclosure.

The sink [source] vertex of the sub-lattice digraph corresponds to a concept in the global context $\mathbb{K}$ iff it has an attribute [object] label. Sink [source] vertices which are concepts are shown in Fig. 8.4c as circles with black [white] fill, while those which are not are represented as point junctions of the arcs from their lower [to their upper] neighbours. Such junctions can be seen, for example, at the top and bottom of the salmon-coloured container in Fig. 8.4c. Each sink [source] vertex is connected by an arc to its counterpart in the parent container. In the case where the former vertex is not a concept, it serves as a collection [distribution] point for a "trunk" arc to [from] its counterpart. These trunk arcs reduce clutter by condensing multiple arcs into a single line.

The CARVE software prototype uses coordinated and multiple views to present the decomposition tree, context bigraph and lattice digraph. The tree view uses a layout typical of file system browsers, which is more space-efficient than that shown in Fig. 8.4b. Selecting a node of the decomposition tree updates the context bigraph and lattice digraph views to show only the corresponding sub-context. These two sub-graphs can be laid out simultaneously to spatially cluster bicliques in the context bigraph and improve the intelligibility of the lattice digraph [61]. By interacting with the decomposition tree, the user can drill down to sub-contexts of interest, for which the context bigraph and lattice digraph can be significantly smaller than for the global context.

8.5.2 User-Guided FCA

The DANCE prototype [58, 60] improves the scalability of FCA for interactive use by allowing the user to steer the analysis towards areas of interest, and to halt construction of the lattice digraph as soon as their analytic objectives are satisfied.

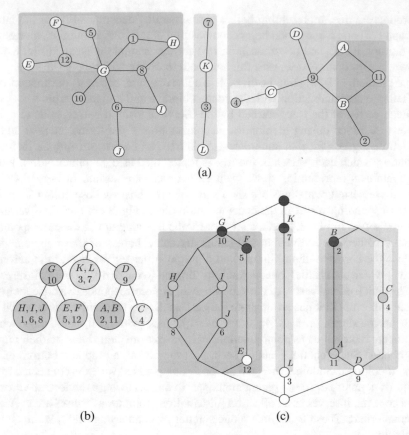

Fig. 8.4: Example context bigraph, the corresponding decomposition tree, and the modified line diagram with discovered sub-lattices within nested containers. (**a**) Context bigraph. (**b**) Decomposition tree. (**c**) Modified line diagram

The resultant lattice digraph will be more task-focused, and, depending on the application, may be considerably smaller, than the lattice digraph for the original context.

We have developed the Dynamic Analysis for Conceptual Exploration (DAɴCE) prototype to explore this possibility, modifying a top-down algorithm for concept enumeration to respond to user control and guidance [60]. Instead of autonomous enumeration of all concepts followed by batch-mode construction of the lattice digraph and corresponding line diagram, DAɴCE allows the user interactive control over the process of concept enumeration and provides dynamic, incremental update of the line diagram. In addition to being able to start, stop, restart and step through the process of concept enumeration, the user can: select a concept of interest and prioritise the enumeration of comparable concepts which are below it in the lattice; or select multiple concepts and prioritise the generation of the concepts which correspond to the set intersections of their intents and extents. Each vertex and arc is displayed either as soon as it is discovered, or in a batch-mode update of the line diagram after a specified number of steps of the enumeration algorithm.

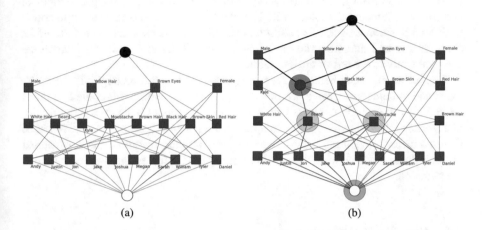

(a)   (b)

Fig. 8.5: The initial line diagram and the result of multiple selection. (**a**) Initial diagram. (**b**) Multiple selection

Visualisation challenges faced by DAɴCE include: ensuring intelligible layout of the partially-constructed diagram; maintaining the user's mental model while vertices and arcs are added; and ensuring that the labelling scheme described in Sect. 8.2.4 applies when some lattice digraph vertices and edges have yet to be discovered. DAɴCE maintains the complete line diagram for the partial order amongst the concepts generated to date, ensuring its consistent interpretation as new concepts are added. Figure 8.5 shows mock-ups of this line diagram for an example formal context consisting of people and their physical attributes. Figure 8.5a depicts the state of the line diagram first presented to the user. By this stage, all of the attribute and object concepts have been generated by the concept enumeration algorithm, labelled, and laid out to establish the framework for insertion of subsequent concepts. Algorithms exist for efficient generation of the requisite Attribute-Object Concept (AOC) poset [12] and for horizontally ordering the atoms and co-atoms to reduce the number of edge crossings in the lattice digraph [61]. Establishing this framework ab initio minimises subsequent disruption of the user's mental model as new concepts are inserted into the line diagram, while the presence and labelling of all attribute and object concepts ensures that intent and extent membership can be read from the outset. Our solution to the problem of efficiently generating only the remaining concepts will be described in a future publication [62].

Figure 8.5b shows the result of the user selecting in this diagram the attribute concepts for "Beard" and "Moustache", which are highlighted in response with light olive halos. This multiple selection triggers the calculation of the extent and intent intersections for the selected concepts. The former corresponds to the infimum, which is accordingly highlighted with a green halo; the latter corresponds to a new concept, which is consequently inserted into the line diagram and highlighted with a purple halo. Since this extent intersection has path length 2 from the supremum, a new row has been inserted to accommodate concepts now with path length 3, and lower neighbours demoted to it. The extent intersection is inserted into row 2 at ordinal position 2 of 5; this position is based on the horizontal barycentre of its associated layer 1 ancestors (co-atoms) and layer 5 descendants (atoms), which are predominantly to the left of the centreline.

The technique described here for user-guided FCA could be applied within the leaf-node containers of CARVE in cases where the corresponding sub-context remains large. This can occur in contexts which are either especially large, or are not particularly amenable to the divide-and-conquer technique employed by CARVE. Whereas CARVE currently performs batch-mode construction and layout of the sub-context digraph during leaf-node traversal, this could be deferred to support user-guided construction.

8.5.3 Structural Navigation

The SORTED prototype [58, 59] supports the retrieval of documents (objects) from a corpus based on queries over the terms (attributes) they contain. User queries are constrained to term combinations which occur in the corpus, and are generalised by removing, or specialised by adding, terms to navigate to comparable concepts. Unlike previous interfaces for structural navigation of the lattice digraph [16, 24, 89], those comparable concepts are not constrained to be neighbours of the current concept. Depicted in Fig. 8.6, the user interface mockup offers valid terms to add or remove from the query. Terms not specified by the user, but which are in the closure of the set of user-specified query terms, are referred to as *closure terms*, and are shown orange in Fig. 8.6. The computational challenge is to compute the set of all concepts reachable from the query concept by the addition or removal of a user-specified query term to facilitate interactive use. Despite structurally navigating the lattice digraph through a keyhole view, the user need not be aware of the lattice digraph's existence, relying instead on intuition established through long-term use of conventional information retrieval interfaces.

In SORTED, FCA provides a mechanism for literal search over the corpus, with the user interface assisting the construction and interactive refinement of conjunctive Boolean queries. The search results are ranked using Latent Semantic Analysis (LSA) according to their cosine similarity to the search terms [46], and the search terms are ranked according to their cosine similarity to the result set. The latter ranking is less conventional, indicating the comparative relevance of the search terms to the result set, from which the user may judge whether the result set is likely to satisfy their requirements. It is these two "semantic" rankings which give rise to the acronym SORTED, which stands for Semantically-Ordered Ranking of Terms and Documents.

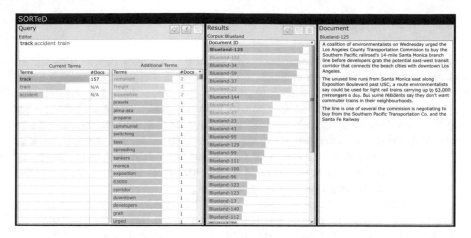

Fig. 8.6: SORTeD interface for information retrieval combining FCA and LSA

In addition to offering *conjunctive* search terms—those which co-occur with the existing search terms—to assist the user to refine the query, the interface also offers, ranks and visually distinguishes *disjunctive* terms—those which are only semantically related to the result set. Conjunctive search terms are shown green in the `Additional Terms` list in Fig. 8.6, while disjunctive search terms are shown blue. Similarly, the `Results` list shows in green the identifiers of documents matching the current conjunctive Boolean query, and in blue the identifiers of those which do not match. In this example, document `Blueland-125` is not a literal match to the conjunctive Boolean query, but is the highest-ranking semantic match.

Selecting a disjunctive term currently initiates a literal query in which the selected term is substituted for the existing set of query terms. A technique has subsequently been described whereby the query is instead edited to include the disjunctive term [59]. The semi-automated editing algorithm searches the lattice digraph constructed from the document corpus for an intent which: contains the disjunctive term; and preserves as many user-specified query terms, and introduces as few new terms, as possible. If more than one intent scores equally against these criteria, the user is asked to choose which of them better reflects their information needs. These choices could be computed for all disjunctive terms while the user assesses the result of their most recent query.

Any closure terms identified during this interactive query refinement process constitute the abbreviated consequent of an implication whose antecedent is the current set of user-specified query terms. Thus while querying the corpus, the user is also discovering implications amongst the terms present in its documents.

8.6 Discovering Insightful Implications

The FCA community has traditionally accepted two different perspectives on the same knowledge, designing methods to represent and navigate the concept lattice and others to manage and infer new implications. Recently, some authors have proposed the use of data visualisation to represent implications and association rules. This issue is far from being fully addressed. In this work we propose a new step to approach the visualisation and exploration of implications. Our starting point is a set of implications, regardless of how it has been extracted from the formal context. We refer the reader interested in algorithms for extracting implications from a formal context to the recent works [3, 55].

This section is organised as follows: Sect. 8.6.1 reviews tools for visualising implications, with particular emphasis on the R package arulesViz [39]. In Sect. 8.6.2 we present a novel model for the visualisation of implications. The model has been guided by semantic characteristics of the attributes, and in particular by the role that each attribute plays in the system of implications.

8.6.1 Visualisation of Implications

We focus here on the implications appearing in a Formal Concept Analysis framework. Concept Explorer[4] [90] provided the ability to both visualise the line diagram and list the associated attribute implications, but user comprehension of the implications was aided by neither tabulation nor data visualisation. In designing the LatViz [2] tool for interactive exploration of concept lattices, Alam et al. [2] considered both rule tabulation and the visualisation of a binary matrix in which the rows correspond to the left-hand sides of implications, the columns to right-hand sides, and unit entries to implications. The use of cell colours was suggested for the latter view to show rule support or lift. However, the authors rejected both views as insufficiently scalable, opting instead for a scatter plot in which each implication was plotted using its support as ordinate and lift as coordinate. Since for implications the product of lift and support is bounded above by 1, each implication x lies below the hyperbola $\mathrm{lift}(x)\,\mathrm{sup}(x) = 1$ on the scatterplot, leading to inefficient use of screen real-estate.

The Cubix FCA-based analytics tool for Business Intelligence [52] included three visualisations to enable progressive exploration of a set of association rules. The first was a matrix view in which each row corresponds to a rule, each column to an attribute-value pair, and cell opacity to rule confidence. The second was a radial drawing of a graph whose vertices represented attribute-value pairs, edges represented co-involvement in a rule, and edge thickness represented rule confidence. Since all implications have the same (100%) confidence, cell opacity and edge thickness in the first and second views, respectively, could for the present purposes be

---

[4] Software available from http://conexp.sourceforge.net.

mapped to other quantities which differ between implications. The third visualisation was a bubble graph in which premises and conclusions were represented as connected bubbles containing the involved attribute-value pairs.

Other tools also display information about rules. Both WiFisViz [47] and VisAR [84], for example, present a very similar approach to the attribute matrix already described in connection with LatViz. The R package arulesViz [37, 38] also supports the visualisation of association rules. It is oriented to present the information so that visual analysis can be further developed. According to Hahsler [37]:

> ... mining association rules often results in a vast number of found rules, leaving the analyst with the task to go through a large set of rules to identify interesting ones. Visualisation and especially interactive visualisation has a long history of making large amounts of data better accessible

arulesViz incorporates the rule visualisation elements used in the tools mentioned above, but it also adds some clustering techniques. It shows the rules in a variety of flexible ways. Detailed information is visualised by means of a scatter plot called a "two-key plot" [86] in which support and confidence are used for the x and y axes and the colour of the points is used to indicate the number of attributes contained in the rule (the order). As a second level to the visualisation, a grouping technique is used to design a new method, called grouped matrix-based visualisation, which is based on a method of clustering rules. A graph-based plot is used to connect the attributes appearing in the association rules.

One of the interesting features of this tool is the incorporation of customisation to perform some kinds of visual analysis. Thus, it is possible to change the information on the axes and display a third or fourth parameter by means of other visualisation elements like colour or shadow. These elements provide a good balance between the simplicity of the plots and quantity of information [28]. We refer the reader to [38] for further details of these features. In the rest of the section, we summarise how arulesViz reveals interesting knowledge.

The starting point is to use the R language to generate the rules [39] using the apriori algorithm.[5] These rules are then directly depicted in a scatter plot (see Fig. 8.7) where each rule is a dot, having two measures on the axes (support and confidence) and a third (lift) visualised using colour intensity.[6] This plot provides significant information about a set of association rules in a very intuitive way. However, it is less suited to visualising implications, for which the value of confidence is always 1. Furthermore, the information displayed is centered on the rule as an atomic element, which must be changed for better knowledge discovery.

In Fig. 8.8, the idea is to show some semantic information: the role played by the attributes in the rules. Thus, the different itemsets in each antecedent and consequent are depicted in the x and y axes respectively. Both index sets are independent, thus

---

[5] As we previously mentioned, we use the extraction method included in arulesViz, but any other method can be used instead.

[6] In this section we are visualising the set of rules of the Adults dataset in the UCI Repository (http://archive.ics.uci.edu/ml/datasets/adult) where the threshold support is set to 0.3 and the confidence to 0.55. This set has 563 rules.

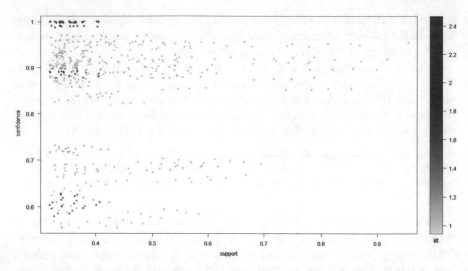

Fig. 8.7: Basic arulesViz scatter plot for 563 association rules

the same number does not represent the same itemset. An interaction feature is included in arulesViz: when a dot is clicked in the plot, the full description of the rule, including the name of the attributes in the LHS and RHS, is shown. Rules can be coloured according to the value of a third feature (in this case, we have selected their lifts). The problem with this plot is that for a huge number of rules, it would be impossible to discover some potentially useful insights. Even for a mid-size volume of rules, the connection between the premise and the conclusion of the rules is not easily identified in the plot.

The third plot included in the arulesViz package is the Two-key plot [86], an example of which is shown in Fig. 8.9. The ordinate and coordinate are the support and confidence, respectively, of a set of association rules. Each rule is represented by a dot, whose colour indicates the number of attributes in the rule. This plot not only shows the usual parameters of the rules (support, confidence, etc.) but it also incorporates some properties that can help to identify the most important rules when a huge number of rules is visualised.

Current trends in data visualisation include some kind of user interaction [79] so that they provide not only a way to communicate the results of data analyses already done, but also to facilitate further exploration by means of visual analysis. In arulesViz, a number of interactive features for scatter plots have been considered, including zooming into a plot, panning around it, and hovering over points to reveal deeper information about the rules they represent. In addition, it also allows rule inspection by selecting a region or a point in the plot, filtering rules, etc. To end the review of arulesViz, we remark that Hahsler and Karpienko [38] recommend that no more than 1000 rules are displayed.

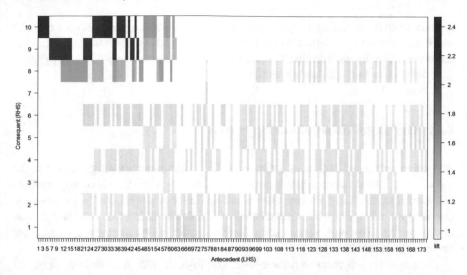

Fig. 8.8: Visualisation of itemset matrix for 563 association rules in arulesViz

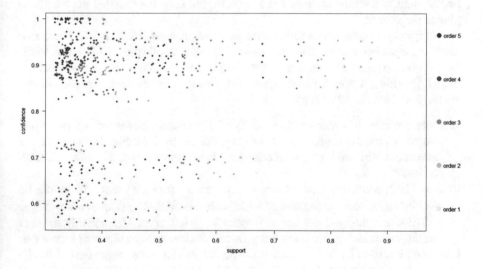

Fig. 8.9: The Two-key plot for 563 association rules in arulesViz

As a general conclusion to this section, in our opinion the visualisation models designed for implications are extremely centered on the implications themselves. As such, they are focussed on displaying the rules and their natural characteristics: support, confidence, lift, and attributes in the RHS or LHS. However, when the number of implications grows, this information is difficult to analyze visually and other elements need to be displayed. In our opinion, these new elements have to be defined so that interesting insights can be made to infer the various roles that the attributes play in the system. This knowledge seems to be difficult to extract from the parameters usually included in the visualisations. Thus, we present a novel approach in this direction in the following section.

8.6.2 Our Data Visualisation Approach

As we have explained, strategies presented in the literature were strongly based on the use of some basic information that could be interesting when the set of implications has a small or medium size. However, when its size grows this information does not provide interesting insights for visual analysis. In this situation, the visualisation has to propose new models based on some semantic information or patterns to discover interesting knowledge.

In this work, we propose a three-step visualisation model to develop a visual analysis process. We approach this model by designing two interactive plots and a final data table. This process will help the user to move from a general set of implications to a smaller subset. The search is guided by the role that the attributes play in the system.

We focus on implications rather than association rules, coopting the basic elements of association rules that also make sense for implications: LHS and RHS. In addition, we have also designed specific parameters to enrich the visualisation model, providing a deeper knowledge of the system. The key elements used in our visual model are the following:

Attribute closure. In order to examine the full semantic power of the premises, we transform the original set Σ of implications by including in the RHS the maximum attribute set for each premise, i.e. we replace each $\mathcal{L} \to \mathcal{R} \in \Sigma$ with $\mathcal{L} \to \mathcal{L}'' \setminus \mathcal{L}$.

RHS and LHS cardinals. Comparison of the cardinalities $|\mathcal{L}|$ and $|\mathcal{R}|$ should be facilitated for each implication $\mathcal{L} \to \mathcal{R}$ with abbreviated consequent $\mathcal{R} = \mathcal{L}'' \setminus \mathcal{L}$. Similar values indicate that an implication is balanced. Alternatively, the user might search for implications where a small $\mathcal{L}$ determines a big $\mathcal{R}$ or vice versa.

Presence in RHS vs LHS. For each attribute, we measure the proportion of implications in which the attribute appears, respectively, on the right and left hand sides.

Global presence. Global presence is defined as the sum of the relative presence in RHS and LHS. In using this, we are searching for interesting attributes instead of interesting rules.

Generator/generated attribute roles. An attribute whose relative presence in the LHS is greater than its relative presence in the RHS will be called a generator. Correspondingly, an attribute with greater presence in the RHS than in the LHS is considered a generated attribute.

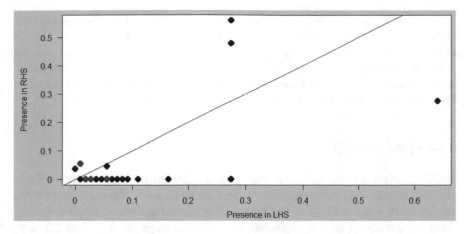

Fig. 8.10: Attribute Plot with generated/generator attribute roles

In the next three subsections, we describe how these parameters are used to design the plots that are the basis of our three step analysis process.

8.6.2.1 Attribute Plot

As we saw in Sect. 8.6.1, most of the relevant literature centers on the visualisation of implications. However, we start our analysis with the attributes, by plotting them according to some semantic information. Particularly, we show the role played by each of them: is it a generator or generated attribute?

Thus, we design a scatterplot in which each attribute, represented by a dot, is plotted using axes representing the frequency with which it appears on the LHS (x axis) and the RHS (y axis) of all implications (see Fig. 8.10). In this plot, we also draw a straight line which represents the identity function. The points located above the line are attributes playing the role of a generated attribute. Conversely, the points below the line are generators. On one hand, a direct interpretation of this plot is the relative presence of the attribute in the premises and abbreviated consequents of the implications. On the other hand, this plot also helps to reveal where the attributes are located in the concept lattice. Top and bottom attributes in the lattice are the attributes placed close to the axes. Thus, the attributes in the lower right quadrant of the diagram appear in closed sets belonging to the lower levels of the concept lattice, whereas the upper left dots correspond to attributes in the closed sets located in the upper levels of the lattice.

In addition, we add some interactive capabilities to this visualisation. The user can click on any point, colouring the attribute red, and triggering a new diagram which will be described in the following subsection. Moreover, in some cases the attributes come from an original many-valued formal context [9] which was converted into a classical one by multiplication of the values. If applicable, we add elements to visualise this variation. So, we consider that all attributes that correspond with different values of an original attribute collectively form a category. When a dot is clicked, all the attributes belonging to the same category are highlighted in blue. As an illustration of this mode, in Fig. 8.10, the red point represents "occupation: transport moving", while the blue ones represent the other occupations: "occupation: Armed forces", "occupation: tech-support" and "occupation: craft repair".

8.6.2.2 Implication Plot

Selection of an attribute in the attribute plot triggers the creation of a second plot showing implications containing the selected attribute. As for the attribute plot, the focus of this visualisation is on the balance between the number of attributes on the RHS and the LHS. This new plot shows the implications gathered by the cardinality of their LHS and RHS, grouping into the same dot all implications which follow the same structural schema. It also adds some additional information like the number of implications in the group and the combination of cardinals on both sides. An example of this diagram is shown in Fig. 8.11. The cardinalities of the LHS and RHS

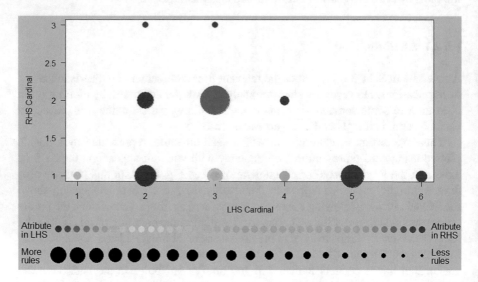

Fig. 8.11: Implication Plot for implications containing the selected attribute

are mapped to the x and y axes respectively. The size of each dot is proportional to the number of implications having the corresponding cardinality pair $(|\mathcal{L}|, |\mathcal{R}|)$. Each dot has an embedded inner point representing the proportion of all implications containing the selected attribute which have the given cardinality pair. Finally, the dot colour represents the main role that the attribute plays in the implications with a given LHS and RHS. Generator and generated roles are highlighted in red and blue, respectively, while intermediate roles are assigned intermediate colours.

Show 10 ▾ entries

LHS	RHS	support	lift	count	right.support	left.support
All	All	.	.	.	All	All
[1] {relationship=Husband,capital-loss=High}	{marital-status=Married-civ-spouse,sex=Male,capital-gain=None}	0.015	2.454	732.000	0.407	0.015
[2] {relationship=Husband,capital-gain=High}	{marital-status=Married-civ-spouse,sex=Male,capital-loss=None}	0.022	2.454	1,082.000	0.407	0.022
[3] {relationship=Husband,capital-gain=Low,hours-per-week=Full-time}	{marital-status=Married-civ-spouse,sex=Male,capital-loss=None}	0.014	2.182	704.000	0.458	0.014
[4] {workclass=Private,relationship=Husband,capital-gain=Low}	{marital-status=Married-civ-spouse,sex=Male,capital-loss=None}	0.017	2.182	835.000	0.458	0.017

Showing 1 to 4 of 4 entries

Fig. 8.12: Properties of the implications associated with the selected attribute

This visualisation reveals whether the attribute plays the generated or generator role in an implication schema, according to their size and LHS and RHS cardinals. Once we have discovered some insights, we can finally get information on all the implications that have been grouped in each dot by clicking on it. This will be shown in a table explained in the following subsection. In future, we plan to add some kind of visual representation to this table, following the models presented in previous works, to help ease exploration as the size of the implication set increases.

8.6.2.3 Rules Data Table

The final step is to show all the implications with the selected LHS and RHS cardinal pair. They are placed in a table with some classical parameters so that the user can examine the set according to the search they performed using the visual tool. This data table has been built by the user in just two actions—the selection of one attribute and one implication schema—allowing the user to develop visual analytics in a few steps. Continuing the example of the previous section, if the points with the biggest RHS are selected—the (2,3) and (3,3) pairs—the implications collected in these points are shown in Fig. 8.12. This table allows the user to sort by a nominated property.

8.7 Conclusions and Future Work

In this chapter we have described the analytical technique of FCA, which derives knowledge in the form of a lattice digraph from big data in the form of a context bigraph. The scalability challenges of construction and interactive visualisation of the resultant DAG have been described, along with three prototype tools we have developed to address aspects of these challenges.

A basis set of attribute implications constitutes an alternative representation of the knowledge discovered by FCA, which has attracted less attention from the visual analytic community than the concept lattice. We have presented a novel visualisation model which assists the user to discover greater insights from implicational bases. Central to this model is identification of the role that attributes play in implications. We have described key elements of the model, and the diagrams and tables used to visually present the information.

Importantly, this chapter has advocated a multiple coordinated views approach which connects visualisations of implications and of the lattice digraph, thereby visually integrating the complementary meanings extracted from a formal context by FCA. Consistent coordination between the views described in this paper, including the context bigraph, lattice digraph, table of basis implications, and data visualisations of the implications, remains an open challenge.

Future work may include the following. Implement the visualisation of individual implications within the lattice digraph (see Fig. 8.3), and investigate extension of this approach to support the pairwise comparison of implications. Explore the use of CARVE to divide and conquer the computation and visualisation of implications. Update the three DST prototypes to implement enhancements described in subsequent publications. In particular, update DANCE to implement the seriation technique described by [61], and integrate the result with CARVE, such that DANCE is used to visualise sub-contexts which are indivisible by CARVE. Publish the description of an algorithm which, given a formal context and its AOC poset, efficiently generates the remaining concepts of the formal context [62], and compare its performance with that of the algorithm currently employed by DANCE. And finally, augment the SORTED prototype to implement the technique described in [59] for maximally preserving an existing query while adding a disjunctive term.

Acknowledgements The contributions to this chapter by the Universidad de Málaga were partially supported by the project TIN2017-89023-P of the Science and Innovation Ministry of Spain, co-funded by the European Regional Development Fund (ERDF).

References

1. Aho A, Garey M, Ullman J (1972) The transitive reduction of a directed graph. SIAM Journal on Computing 1(2):131–137

2. Alam M, Le TNN, Napoli A (2016) LatViz: A new practical tool for performing interactive exploration over concept lattices. In: Huchard M, Kuznetsov S (eds) Proc. 13th Intl. Conf. CLA, CEUR-WS.org, vol 1624, pp 9–20

3. Andrews S (2009) In-Close, a fast algorithm for computing formal concepts. In: Rudolph S, Dau F, Kuznetsov S (eds) Supplementary Proc. 17th Intl. Conf. Conceptual Structures, CEUR-WS.org, vol 483

4. Andrews S, Hirsch L (2016) A tool for creating and visualising formal concept trees. In: Andrews S, Polovina S (eds) Proc. 5th Conceptual Structures Tools & Interoperability Workshop (CSTIW 2016) held at the 22nd Intl. Conf. Conceptual Structures (ICCS 2016), CEUR-WS.org, vol 1637, pp 1–9

5. Archambault D, Munzner T, Auber D (2008) Grouseflocks: Steerable exploration of graph hierarchy space. IEEE Transactions on Visualization and Computer Graphics 14(4):900–913

4. Armstrong WW (1974) Dependency structures of database relationships. Information Processing 74:580–583

7. Baklouti F, Lévy G, Emilion R (2005) A fast and general algorithm for Galois lattices building. Journal of Symbolic Data Analysis 2:19–31

8. Belohlávek R, Snasel V (eds) (2005) Proc. 2005 Intl. Workshop CLA, vol 162, CEUR-WS.org, http://ceur-ws.org/Vol-162/

9. Bělohlávek R, Vychodil V (2005) What is a fuzzy concept lattice? In: [8], pp 34–45

10. Benito-Picazo F, Cordero P, Enciso M, Mora Á (2019) Minimal generators, an affordable approach by means of massive computation. J Supercomput 75(3):1350–1367. https://doi.org/10.1007/s11227-018-2453-z

11. Berry A, Pogorelcnik R, Sigayret A (2011) Vertical decomposition of a lattice using clique separators. In: Napoli A, Vychodil V (eds) Proc. 2011 Intl. Conf. CLA, CEUR-WS.org, vol 959, pp 15–29

12. Berry A, Gutierrez A, Huchard M, Napoli A, Sigayret A (2014) Hermes: a simple and efficient algorithm for building the AOC-poset of a binary relation. Annals of Mathematics and Artificial Intelligence 72(1):45–71

13. Bhatti MU, Anquetil N, Huchard M, Ducasse S (2012) A catalog of patterns for concept lattice interpretation in software reengineering. In: Proc. 24th Intl. Conf. Software Eng. & Knowledge Eng. (SEKE'2012), Knowledge Systems Institute Graduate School, pp 118–124

14. Boukhelifa N, Roberts JC, Rodgers P (2003) A coordination model for exploratory multiview visualization. In: Proc. Intl. Conf. Coordinated and Multiple Views in Exploratory Visualization (CMV 2003), IEEE, pp 76–85

15. Card S, Mackinlay J, Shneiderman B (1999) Readings in information visualization: Using vision to think. Morgan Kaufmann

16. Carpineto C, Romano G (2004) Exploiting the potential of concept lattices for information retrieval with CREDO. J Univers Comput Sci 10(8):985–1013. https://doi.org/10.3217/jucs-010-08

17. Chen CP, Zhang CY (2014) Data-intensive applications, challenges, techniques and technologies: A survey on big data. Inf Sci 275:314–347

18. Choi V, Huang Y (2006) Faster algorithms for constructing a Galois lattice, enumerating all maximal bipartite cliques and closed frequent sets. Presented at SIAM Conf. Discrete Mathematics

19. Clémençon S, Arazoza HD, Rossi F, Tran VC (2011) Hierarchical clustering for graph visualization. In: Proc. XIXth European Symposium on Artificial Neural Networks (ESANN 2011), i6doc.com, pp 227–232

20. Cordero P, Enciso M, Mora Á, de Guzmán IP (2002) SLFD logic: Elimination of data redundancy in knowledge representation. In: Advances in Artificial Intelligence – IBERAMIA 2002, pp 141–150. https://doi.org/10.1007/3-540-36131-6

21. Cordero P, Enciso M, Mora Á, Ojeda-Aciego M (2012) Computing minimal generators from implications: a logic-guided approach. In: Proc. 2012 Intl. Conf. CLA, CEUR-WS.org, vol 972, pp 187–198
22. Davey BA, Priestley HA (2002) Introduction to Lattices and Order, 2nd edn. Cambridge University Press, England
23. Di Battista G, Eades P, Tamassia R, Tollis I (1999) Graph Drawing: Algorithms for the visualization of graphs. Prentice Hall, NJ, USA
24. Ducrou J, Eklund P (2006) Browsing and searching MPEG-7 images using Formal Concept Analysis. In: Proc. 24th IASTED Intl. Conf. Artificial Intelligence and Applications (AIA'06), Innsbruck, Austria, pp 317–322
25. Eklund P (ed) (2004) Proc. 2nd Intl. Conf. FCA, LNCS, vol 2961, Springer. https://doi.org/10.1007/b95548
26. Eklund P, Villerd J (2010) A survey of hybrid representations of concept lattices in conceptual knowledge processing. In: Kwuida L, Sertkaya B (eds) Proc. 2010 Intl. Conf. FCA, pp 296–311. https://doi.org/10.1007/978-3-642-11928-6
27. Eklund PW, Ducrou J, Brawn P (2004) Concept lattices for information visualization: Can novices read line diagrams? In: [25], pp 57–73
28. Few S (2012) Show Me the Numbers: Designing Tables and Graphs to Enlighten, 2nd edn. Analytics Press
29. Freese R (2004) Automated lattice drawing. In: [25], pp 112–123
30. Ganter B (2004) Conflict avoidance in additive order diagrams. J Univers Comput Sci 10(8):955–966. https://doi.org/10.3217/jucs-010-08
31. Ganter B, Obiedkov S (2016) Attribute exploration, Springer, chap 4, pp 125–185
32. Ganter B, Wille R (1999) Formal Concept Analysis: Mathematical Foundations. Springer
33. Gaume B, Navarro E, Prade H (2010) A parallel between extended Formal Concept Analysis and bipartite graphs analysis. In: Proc. IPMU'10, Berlin, Heidelberg, LNCS, vol 6178, pp 270–280
34. Godin R, Saunders E, Jescei J (1986) Lattice model of browsable data spaces. Information Sciences 40(2):89–116
35. Grossman RL (2018) A framework for evaluating the analytic maturity of an organization. Intl J Information Management 38(1):45–51. https://doi.org/10.1016/j.ijinfomgt.2017.08.005
36. Guigues Jl, Duquenne V (1986) Familles minimales d'implications informatives d'un tableau de données binaires. Mathématiques et Sciences Humaines 95:5–18
37. Hahsler M (2017) arulesViz: Interactive visualization of association rules with R. The R Journal 9(2):163–175
38. Hahsler M, Karpienko R (2017) Visualizing association rules in hierarchical groups. J Bus Econ 87:317–335
39. Hahsler M, Grun B, Hornik K (2005) arules - a computational environment for mining association rules and frequent item sets. J Stat Software 14:1–25
40. Hannan T, Pogel A (2006) Spring-based lattice drawing highlighting conceptual similarity. In: Proc. 4th Intl. Conf. FCA, LNCS, vol 3874, pp 264–279
41. Herman I, Melançon G, Marshall MS (2000) Graph visualization and navigation in information visualization: a survey. IEEE Transactions on Visualization and Computer Graphics 6:24–43
42. Ignatov DI, Nourine L (eds) (2018) Proc. 2018 Intl. Conf. CLA, vol 2123, CEUR-WS.org
43. Kriegel F (2013) Visualization of conceptual data with methods of Formal Concept Analysis. PhD thesis, Technische Universitat Dresden
44. Kuznetsov SO, Obiedkov SA (2001) Algorithms for the construction of concept lattices and their diagram graphs. In: [70], pp 289–300
45. Lamping J, Rao R, Pirolli P (1995) A focus+context technique based on hyperbolic geometry for visualizing large hierarchies. In: Proc. SIGCHI Conf. Human Factors in Computing Systems, New York, NY, USA, pp 401–408

46. Landauer T, McNamara D, Dennis S, Kintsch W (eds) (2013) Handbook of Latent Semantic Analysis. Taylor & Francis
47. Leung KS, Leung CKS, P Irani P, L Carmichael C (2008) WiFIsViz: Effective visualization of frequent itemsets. IEEE Intl Conf Data Mining pp 875–880
48. Liiv I (2008) Pattern discovery using seriation and matrix reordering: A unified view, extensions and an application to inventory management. PhD thesis, Tallinn University Of Technology
49. Lindig C (2000) Fast concept analysis. In: Working With Conceptual Structures - Contributions to ICCS 2000, pp 152–161
50. Maddouri M, Gammoudi J (2007) On semantic properties of interestingness measures for extracting rules from data. In: Adaptive and Natural Computing Algorithms, Springer, LNCS, vol 4431, pp 148–158
51. Melo C, Le-Grand B, Aufaure MA, Bezerianos A (2011) Extracting and visualising tree-like structures from concept lattices. In: Proc. 15th Intl. Conf. Information Visualisation, IEEE Computer Society, Washington, DC, USA, pp 261–266. https://doi.org/10.1109/IV.2011.46
52. Melo C, Mikheev A, Le-Grand B, Aufaure M (2012) Cubix: A visual analytics tool for conceptual and semantic data. In: Vreeken J, Ling C, Zaki MJ, Siebes A, Yu JX, Goethals B, Webb G, Wu X (eds) 12th International Conference on Data Mining Workshops, IEEE, pp 894–897. https://doi.org/10.1109/ICDMW30884.2012.41
53. Mora Á, Cordero P, Enciso M, Fortes I, Aguilera G (2012) Closure via functional dependence simplification. Int J Comput Math 89(4):510–526
54. Nauer E, Toussaint Y (2009) CreChainDo: An iterative and interactive web information retrieval system based on lattices. Intl J General Systems 38(4):363–378
55. Outrata J, Vychodil V (2012) Fast algorithm for computing fixpoints of Galois connections induced by object-attribute relational data. Information Sciences 185(1):114–127
56. Owais S, Gajdoš P, Snášel V (2005) Usage of genetic algorithm for lattice drawing. In: [8], pp 82–91
57. Papoulis A (1984) Probability, Random Variables, and Stochastic Processes, 2nd edn. McGraw-Hill Series in Electr. Eng., McGraw-Hill
58. Pattison T (2014) Interactive visualisation of formal concept lattices. In: Burton J, Stapleton G, Klein K (eds) Joint Proc. 4th Intl. Workshop Euler Diagrams and 1st Intl. Workshop Graph Visualization in Practice, CEUR-WS.org, vol 1244, pp 78–79
59. Pattison T (2018) Interactive query refinement using Formal Concept Analysis. In: [42], pp 207–218
60. Pattison T, Ceglar A (2014) Interaction challenges for the dynamic construction of partially-ordered sets. In: Proc. 2014 Intl. Conf. CLA, CEUR-WS.org, vol 1252, pp 23–34
61. Pattison T, Ceglar A (2019) Simultaneous, polynomial-time layout of context bigraph and lattice digraph. In: Proc. 15th Intl. Conf. FCA, pp 223–240
62. Pattison T, Ceglar A (2021) Towards interactive transition from AOC poset to concept lattice. In: 2021 Intl. Conf. FCA, to be submitted
63. Pattison T, Phillips M (2001) View coordination architecture for information visualisation. In: Eades P, Pattison T (eds) Proc. Australian Symposium on Information Visualisation, ACS
64. Pattison T, Weber D, Ceglar A (2014) Enhancing layout and interaction in Formal Concept Analysis. In: IEEE Pacific Visualization Symposium (PacificVis), pp 248–252
65. Pattison T, Ceglar A, Weber D (2018) Efficient Formal Concept Analysis through recursive context partitioning. In: [42], pp 219–230
66. Poelmans J, Elzinga P, Viaene S, Dedene G (2010) Formal Concept Analysis in knowledge discovery: A survey. In: Conceptual Structures: From Information to Intelligence, LNCS, vol 6208, Springer, pp 139–153
67. Poelmans J, Kuznetsov SO, Ignatov DI, Dedene G (2013) Formal Concept Analysis in knowledge processing: a survey on models and techniques. Expert Systems with Applications 40:6601–6623. https://doi.org/10.1016/j.eswa.2013.05.007

68. Priss U (2006) Formal Concept Analysis in Information Science. Annual Review of Information Science and Technology 40:521–543
69. Pulo K, Eades P, Takatsuko M (2003) Smooth structural zooming of h-v inclusion tree layouts. In: Proc. Intl. Conf. Coordinated & Multiple Views in Exploratory Visualization, pp 14–25. https://doi.org/10.1109/CMV.2003.1214999
70. Raedt LD, Siebes A (eds) (2001) PKDD '01: Proc. 5th European Conf. Principles of Data Mining and Knowledge Discovery, Springer. https://doi.org/10.1007/3-540-44794-6
71. Rival I (1993) Reading, drawing, and order. In: Rosenberg IG, Sabidussi G (eds) Algebras and Orders, Springer, Dordrecht, Netherlands, pp 359–404
72. Roberts JC (2007) State of the art: Coordinated and multiple views in exploratory visualization. In: Proc. 5th Intl. Conf. Coordinated and Multiple Views in Exploratory Visualization (CMV2007), IEEE, pp 61–71
73. Rodríguez-Lorenzo E, Adaricheva K, Cordero P, Enciso M, Mora Á (2017) Formation of the d-basis from implicational systems using simplification logic. Intl J General Systems 46(5):547–568
74. Rodríguez-Lorenzo E, Bertet K, Cordero P, Enciso M, Mora Á (2018) Direct-optimal basis computation by means of the fusion of simplification rules. Discrete Applied Math 249:106–119
75. Sertkaya B (2010) A survey on how description logic ontologies benefit from FCA. In: Proc. 2010 Intl. Conf. CLA, CEUR-WS.org, vol 672, pp 2–21
76. Shaaban E, Helmy Y, Khedr A, Nasr M (2012) Business intelligence maturity models: Toward new integrated model. In: Proc. Intl. Arab Conf. Information Technology, pp 276–284
77. Shields C (ed) (2012) The Oxford Handbook of Aristotle. Oxford Univ. Press
78. Shneiderman B (1992) Tree visualization with treemaps: a 2-D space-filling approach. ACM Transactions on Graphics 11(1):92–99
79. Simon P (2014) The Visual Organization: Data Visualization, Big Data, and the Quest for Better Decisions. Harvard Business Review
80. Stewart I (2004) Galois theory, 3rd edn. Chapman & Hall/CRC Mathematics
81. Stumme G, Taouil R, Bastide Y, Pasquier N, Lakhal L (2002) Computing Iceberg concept lattices with Titanic. Data & Knowledge Eng 42(2):189–222
82. Sucharitha V, Subash SR, Prakash P (2014) Visualization of big data: Its tools and challenges. Intl J Applied Eng Res 9(18):5277–5290
83. Sugiyama K, Tagawa S, Toda M (1981) Methods for visual understanding of hierarchical system structures. IEEE Trans Systems, Man and Cybernetics 11(2):109–125
84. Techapichetvanich K, Datta A (2008) VisAR: A new technique for visualizing mined association rules. In: Proc. 1st Intl. Conf. Advanced Data Mining and Applications, Springer, LNAI, vol 3584, pp 88–95
85. Thomas JJ, Cook KA (eds) (2005) Illuminating the Path: The Research and Development Agenda for Visual Analytics. IEEE Press
86. Unwin A, Hofmann H, Bernt K (2001) The TwoKey plot for multiple association rules control. In: [70], pp 472–483
87. Wild M (2017) The joy of implications, aka pure Horn functions: mainly a survey. Theor Comput Sci 658(B):264–292. https://doi.org/10.1016/j.tcs.2016.03.018
88. Wille R (1982) Restructuring lattice theory: An approach based on hierarchies of concepts. In: Rival I (ed) Ordered sets, Reidel Publishing, Dordrecht–Boston, pp 445–470
89. Wray T, Eklund P (2011) Exploring the information space of cultural collections. In: Valtchev P, Jäschke R (eds) Proc. 9th Intl. Conf. FCA, Springer
90. Yevtushenko SA (2000) Concept Explorer: a system for data analysis. In: Proc. 7th National Conference on Artificial Intelligence KII-2000, Russia, pp 127–134, in Russian
91. Yevtushenko SA (2004) Computing and visualizing concept lattices. PhD thesis, Technischen Universität Darmstadt
92. Zschalig C (2007) An FDP-algorithm for drawing lattices. In: Proc. 2007 Intl. Conf. CLA, CEUR-WS.org, vol 331, pp 58–71

Chapter 9
Formal Methods in FCA and Big Data

Domingo López-Rodríguez, Emilio Muñoz-Velasco, and Manuel Ojeda-Aciego

9.1 Introduction

The term *Big Data* generally refers to massive quantities of data exceeding the typical processing and computing capacity of conventional databases and data analysis techniques. Due to its particular features, the *Big Data* problem requires the design of *ad hoc* tools and methods to analyze and extract patterns from large-scale data. Increased data storage capabilities and processing power, together with the availability of massive volumes of data constitute the cause of the recent rise of Big Data. Organizations have more data available than they can process since, in general, their computing resources and technologies are limited and not adapted to the large-scale processing inherent in Big Data. In addition to the obvious massive data volume, Big Data has associated other specific qualities, often referred to as the five Vs: Volume, Variety, Velocity, Veracity and Value (the worth of the information extracted from data) [36, 39].

Machine learning is a subdomain of computer science used to analyze data, automating the construction of analytical models. The purpose of machine learning algorithms is to learn from existing data without the need to explicitly program an analytical model. The trained models learn from preceding data and calculations and aim to produce certain and replicable decisions and outcomes.

Machine learning has an extensive variety of applications in fields such as artificial intelligence, optimal control, statistics, information theory, optimization theory, and many other disciplines of mathematics, engineering, and science [66].

The essence of Big Data Analytics is mining and extracting *patterns* for decision-making, prediction and other types of inference, from massive data. A significant principle is that the extracted patterns should be meaningful and provide some understanding about the analyzed data. As machine learning models are used in critical and sensitive areas like medicine, the criminal justice system, and financial

D. López-Rodríguez (✉) • E. Muñoz-Velasco • M. Ojeda-Aciego
Universidad de Málaga, Departamento Matemática Aplicada, Málaga, Spain
e-mail: dominlopez@uma.es; ejmunoz@uma.es; aciego@uma.es

R. Missaoui et al. (eds.), *Complex Data Analytics with Formal Concept Analysis*,
https://doi.org/10.1007/978-3-030-93278-7_9

markets, the inability of humans to understand these patterns becomes problematic [22, 46].

In this sense, logic-based approaches are perfect candidates to help make machine learning models *understandable*, be it through the hybridization with other techniques [71], or as the core of an expert system [25], just to mention two interesting cases.

Formal Concept Analysis (FCA) is a formal framework, built on lattice theory and Galois connections, allowing to mathematically formalize the notion of "concept" (a general idea that corresponds to some type of entity and that can be characterized by some essential features of the class). The mechanisms used to extract these concepts from a dataset in FCA allow us to hierarchically organize them in the so-called "concept lattice". An important aspect to emphasize is that the concept lattice captures *all* the implicit knowledge that can be deduced from a formal context. Another way to extract knowledge from a context is in the form of implications. In essence, an "attribute implication" is an expression of type $A \to B$ where A and B are sets of attributes/properties, and we say that this is fulfilled if every object that has the attributes of A has those of B as well. The use of implicational systems allows to handle all this implicit knowledge (knowledge that is tacit in the experiences, but not codified nor formalized [65]) and to reason over it. We believe that this logic-based approach is more suitable to provide understandable answers and, hence help avoid the lack of interpretability and explainability of the results.

Explainability and interpretability are often used interchangeably. Although they are very closely related, a greater comprehension of the Big Data problems can be gained if their differences are well-understood.

There is no mathematical definition of interpretability. An approximation could be [56]: "Interpretability is the degree to which a human can understand the cause of a decision". Another one is: "Interpretability is the degree to which a human can consistently predict the model's result" [47]. A higher interpretability of a model means an easier understanding of why certain decisions or predictions are made. Interpretability is about the capability of predicting the outcome given a change in the input data or in the algorithm parameters. That is, it is the ability to understand the relationship between the input provided and the outcome given by the model.

Explainability, meanwhile, is the ability to explain the internal mechanics of a system in human terms. The difference with interpretability is very subtle, and this makes the two terms frequently interchanged. However, one can say, in other words, that interpretability is the ability to understand the influence of the mechanics, and explainability is the ability to describe its mechanics.

Formal logical methods, such as the management of implicational systems stated above, are, contrary to the statistical techniques already in use, highly interpretable and explainable, what makes them more suitable for reasoning and extracting/representing knowledge.

The main issue is to what extent these logic-based methods can be applied to a Big Data framework. The scalability of FCA methods and algorithms depends on, and is constrained by, the complexity of the problems to be solved, namely: the identification of concepts and construction of the concept lattice from a *massive* context, the computation of a *canonical basis* of implications and the efficient computation of closures with respect to an implication system.

Here, the term *basis* is used for a set of implications, which, besides being sound and complete, satisfies some minimality condition among all equivalent sets of implications (defining the same *closure* system); thus, there may be different types of bases. The *canonical* basis (also called *stem* basis or Duquenne-Guigues' basis [41]) is a basis of minimal cardinality.

It has been proved [7, 33] that, unless $P = NP$, there is no polynomial delay algorithm to enumerate the implications of the canonical basis, both in the lectic and in the reverse order. Let us note that, even in the simplest cases, the canonical basis can have exponential size [51]. More precisely, the problems of enumerating all pseudo-intents in a formal context (premises of implications in the canonical basis) and computing the lectically largest pseudo-intent belongs to the coNP-complete complexity class [7]. Even if the enumeration of pseudo-intents can be done without an explicit ordering, this result is still true, as it was proved by comparing this problem with that of enumerating minimal transversals in a hypergraph [34, 45]. Thus, somehow taming the complexity issues by designing alternative approaches or increasing the threshold up to which the complexity is still tractable is a major open question.

Therefore, most efforts are focused on two (overlapping) strategies: the development of algorithms which, in the average case, have a short running time and low complexity, and the definition of probabilistic and approximated logic approaches, which may capture essential knowledge with high probability and reduced computational effort. These two strategies seek to build better techniques to explore large datasets.

The objective of this work is to present a survey of the theoretical and technical foundations of some trends of FCA with respect to the issues stated above. We have explored and collected formal and technical developments regarding three pillars:

- The efficient construction of the concept lattice associated to a formal context by the use of simplification procedures.
- The definition and computation of bases of implications different from the canonical basis, satisfying other optimality conditions, and of logic rules and operators that may lead to more efficient inference systems.
- The definition of *probably approximately correct implication bases*, as a way to capture most of the knowledge contained in a dataset with efficient and scalable algorithms.

The remainder of this work is structured as follows: in Sect. 9.2, we present techniques for context and concept lattice reduction, including methods to simplify the structure and decrease the practical complexity of the problem. In Sect. 9.3, logic tools are described in order to operate on implication sets and define optimal bases which may have an important impact on the performance of inference systems. In Sect. 9.4, minimal generators are presented as a means to encapsulate a compressed representation of the knowledge present in a formal context. Finally, Sect. 9.5 presents the foundations for approximated implication bases and efficient algorithms that can be used to compute them. We finish with Sect. 9.6, where we present some conclusions and potential future trends in the application of formal methods to Big Data.

9.2 Context and Concept Lattice Reduction Methods

As stated before, the complexity of (the algebraic and logic tools related to) FCA can be considered one of the most outstanding problems when trying to make effective use of FCA in Big Data situations. Thus, some previous steps may be taken in order to decrease the computational efforts needed to build the concept lattice and to extract the implicational basis, actually reducing the complexity of the context or of the associated concept lattice, both in terms of magnitude (size of object and attributes sets) and of inter-relationships, while relevant information is kept. This set of techniques is known as *concept lattice reduction methods*.

In [31], the authors make a classification of concept lattice reduction methods into three categories: redundant information removal, concept lattice simplification and attribute selection.

Redundant Information Removal

The aim of redundant information removal techniques is, given the formal context $\mathbb{K}$, to build a formal context $\mathbb{K}'$ whose concept lattice has the same structure as that of $\mathbb{K}$.

Definition 9.1 For a given formal context $\mathbb{K} = (G, M, I)$, an object $g \in G$ (or an attribute $m \in M$ or an incidence $i \in I$, resp.) is said to be *redundant information* if its removal builds a formal context $\mathbb{K}' = (G', M', I')$ with $G' = G \smallsetminus \{g\}$ (resp., $M' = M \smallsetminus \{m\}$, or $I' = I \smallsetminus \{i\}$) whose concept lattice is isomorphic to that of $\mathbb{K}$.

Note that redundant information is a term coined from the purely algebraic point of view, but not from the point of view of the application domain. This means that not all the redundant information in a context may be considered not relevant, since the relevance of an object or attribute is defined by the application context.

The *clarification* of a context [38] consists in replacing a set of objects $\{g_i\} \subseteq G$ with exactly the same attributes by a representative object (that is, substitute all g_i by a single object $g \in [g_i]$, where $[g_i]$ denotes the equivalence class of the objects with the same attributes as g_i) and making the analogous substitution in attributes. This way, the new formal context $\mathbb{K}' = (G', M', I|_{G' \times M'})$, isomorphic to $\mathbb{K}$, is obtained by removing duplicated rows and columns of the formal context.

Depending on the background knowledge, it may be advisable not to merge two objects with the same attributes. In the application domain, they may still be considered different from one another and changing a concept extent can influence the quality measures [52] computed from it. Thus, it is recommended to proceed carefully when performing *clarification* in some contexts.

Other research lines [73] aim to analyse the information in the incidence table in terms of the partitions induced on the sets of objects and attributes by some functions of single attributes and objects of the context, obtaining the so-called Formal Equivalence Analysis (FEΛ). Rather than looking on the effect of these partitions on set representation, as done in Rough Sets Theory [63], the emphasis is put on making explicit the information in the context.

Another kind of reduction is the removal of attributes representable by a combination of other attributes [38]. The former are called *reducible* attributes:

Definition 9.2 Given a formal context $\mathbb{K} = (G, M, I)$, an attribute $m \in M$ is called *reducible* if there exists $M' \subseteq M$ with $m \notin M'$ such that the extent of m coincides with the extent of M'. In other words, m is reducible if it is contained in the closure of M'.

Analogously, an object $g \in G$ is *reducible* if there is a $G' \subseteq G$ with $g \notin G'$ such that the intents of g and G' coincide. In the dual context $\tilde{\mathbb{K}} = (M, G, \tilde{I})$ ($m\tilde{I}g \iff gIm$), the closure of g and that of G' (as attribute sets) are equal.

The result of processing a formal context by using clarification and the removal of reducible attributes gives the *standard context*, and it is isomorphic to the original context [38]. The complexity of constructing the standard context using the clarification and reduction methods is $O(|G||M|^2)$, hence polynomial.

These *clarification* and *reduction* methods are only a sample of the techniques which aim to minimize the number of attributes in the context while maintaining the isomorphic correspondence between the original concept lattice and the associated to the reduced context. In general, these techniques focus on computing the possible *reducts* of a context:

Definition 9.3 [48] Given a context $\mathbb{K} = (G, M, I)$, a set $M' \subsetneq M$ is called *consistent* if the concept lattice associated to $(G, M', I|_{G \times M'})$ is isomorphic to that of $\mathbb{K}$. A *reduct* is a minimal consistent set of attributes, that is, X is a reduct if X is consistent and no $X' \subsetneq X$ is consistent.

An equivalent definition of *reduct* is that it is a maximal subset of M without reducible attributes [49].

Although the *clarification* and *reduction* methods have polynomial complexity (in the number of attributes), the general problem of computing all reducts of a formal context has an exponential computational cost, since the number of possible subsets to explore is exponential.

In order to propose a mechanism to compute a minimal set of attributes, in [76], the concept of discernibility matrix is introduced:

Definition 9.4 Given a formal context $\mathbb{K}$, the *discernibility* between concepts (A, B) and (C, D) is defined by the symmetric difference between sets B and D:

$$\text{dis}\,((A, B), (C, D)) = (B \smallsetminus D) \cup (D \smallsetminus B)$$

The *discernibility matrix* associated to the context is a matrix indexed by the concepts, whose entries are the corresponding pairwise discernibilities. The set of non-empty elements of the discernibility matrix is represented by $\Lambda_{\mathbb{K}}$.

Using a discernibility matrix, all the reducts can be obtained, by the application of the following result:

Theorem 9.1 *Given a context* $\mathbb{K} = (G, M, I)$ *and* $\Lambda_{\mathbb{K}}$ *the set of non-empty elements of its discernibility matrix, then* $D \subseteq M$ *is consistent if and only if* $D \cap H \neq \emptyset$ *for all* $H \in \Lambda_{\mathbb{K}}$.

This theorem shows that to find a reduct of a formal context is to find the minimal subset D of attributes which verify the mentioned restriction.

In the same work [76], the authors classify the attributes as *absolutely necessary*, *relatively necessary* or *absolutely unnecessary* to reflect whether they are necessary, contingent or superfluous.

Definition 9.5 Given a formal context $\mathbb{K} = (G, M, I)$, an attribute is called absolutely necessary if it is present in all reducts; it is said to be relatively necessary if it is present in at least one, but not in all minimal consistent sets; and finally, an attribute is unnecessary if it is not in any reduct.

However, the computation of discernibility the matrix requires a high computational effort, since all pairs of concepts are checked for its construction.

A more adequate refinement is proposed in [67], where the authors determine that many of the discernibility sets computed following the strategy in [76] do not actually contribute to finding the reducts. It was proved that the only discernibility sets that need to be computed are those related to adjacent concepts in the concept lattice, that is, to pairs of concept-superconcept. This number of computations corresponds to the number of edges in the lattice, and is clearly lower than the total number of concept pairs. However, both discernibility matrix methods need to compute all formal concepts beforehand, which requires an exponential time in the worst-case.

In [48], a comparison between the traditional clarification and reduction method with that of the discernibility matrix is presented. It is proved that the sets of attributes that are merged in the clarification and reduction steps are exactly minimal non-empty discernibility sets, therefore with the clarification and reduction we can obtain the same result as with methods of attribute reduction based on the discernibility matrix. The relevance of this result can be better understood if we see that the complexity of the clarification and reduction method is $O(|G||M|^2)$ (polynomial) and the discernibility matrix requires $O(|M|^2|\mathcal{L}|))$, where $\mathcal{L}$ is the set of concepts, which can be of cardinality up to $2^{\min\{|G|,|M|\}}$, so actually the computation of the discernibility matrix is exponential in the worst-case.

A further improvement is presented in [49], where a modified algorithm is introduced that only computes the minimal discernibility sets, allowing for polynomial time complexity, in contrast to the exponential complexity mentioned above. In addition, it is stated that if the consistent elimination of all unnecessary computations of discernibility sets is pursued, the resulting method is just cosmetically different from clarification and reduction. In conclusion, the methods based on the computation of the discernibility matrix cannot become more efficient than those based on clarification and reduction of the context.

It is worth noting that the classification into three categories (necessary, contingent, superfluous) is independent of the following frameworks: (usual) concept lattices, property-oriented concept lattices and object-oriented concept lattices [55]. Thus, only one of these three types of lattices needs to be considered when computing the reducts, and the algorithms developed, such as [15, 74], can be applied to any of them.

A different reduction strategy is studied in [44]. It is focused on the incidence relationship, considering the influence of eliminating a single incidence from the formal context in the complexity of building a the new reduced concept lattice.

Simplification of the Structure

Other reduction techniques apply an abstraction of the concept lattice, looking for a high-level overview that preserves only the essential aspects. This type of techniques are grouped as *simplification methods*. In general, these approaches try to build a simplified context or lattice, by grouping similar objects, attributes or concepts.

Another line of research is the use of matrix factorization techniques which allow to decompose the formal context into a simpler representation. Most works [23, 24, 37] in this line use the technique of singular value decomposition (SVD), which allows to project a high-dimensional matrix into one of lower dimensionality. The SVD is used to embed objects into an Euclidean space where similarity measures can be defined, such as the cosine similarity between the vector representation of objects [23], or induce equivalence classes of objects or attributes.

Another application of matrix factorization, is to find key factors which could represent, exactly or approximately, an underlying structure in data. The linear combination of these factors is a lossy approximation of the original formal context. Since the number of factors is usually much lower than the cardinality of the set of objects, they can be considered as representative rows in the formal context, achieving a significant reduction and preserving (with a little loss) all the information contained in the context. Among matrix factorization techniques for context reduction, we can find non-negative matrix factorization [50] and binary matrix factorization [13].

In other works, the matrix factorization is induced not from the formal context, but from the concept lattice itself [11, 62], and the aim of those methods is to get the best representation of the knowledge as a set of factors, which are then called *concept factors* [13].

Selection of Attributes and Concepts

The underlying idea is the application of some criterion which quantifies the importance of attributes and concepts, in order to keep a subset of those with the highest relevance. The importance criterion is application-dependent and, thus, varies between different domains.

There are two different approaches in the selection of attributes and concepts: select relevant items *a posteriori*, after the computation of the whole concept lattice, or make an *a priori* analysis of a relevance measure and then build the *reduced* concept lattice with pre-chosen constraints.

As commented above, the first strategy makes use of the complete concept lattice to infer the importance of attributes, objects and concepts.

A simple notion of relevance may be related to the cardinality of the intension or extension of a concept, but more refined definitions include the assignment of weights [10] to each attribute, and then selecting formal concepts considered relevant. In this case, the relevance of a concept is calculated as the average relevance of the attributes in its intension [75]. In [10], it is also proposed the use of minimal generators of concepts, instead of their intension, to evaluate the importance of a formal concept.

A comprehensive review of interestingness measures of concepts is presented in [52]. In that work, the measures considered are compared regarding aspects such as efficiency of computation and applicability to noisy data.

From the algorithmic perspective, a major issue in this strategy is that concepts must be computed before verifying whether their relevance is above a predefined threshold, due to the inherent complexity of enumerating all formal concepts.

As a solution to this problem, another research line attempts to pre-select attributes based on a relevance measure (thus, used *a priori*) and therefore define constraints that the computed concepts must fulfill.

The relationship between the notions of frequent formal concept (a concept whose support is above a predefined threshold) and of (the analogous) frequent itemsets in transaction databases was exploited in [72] to introduce the Titanic algorithm for the construction of *iceberg lattices*. An *iceberg* lattice consists of frequent itemsets associated to a predefined minimum support, and is therefore more efficient to build than the complete concept lattice.

Another method to reduce the complexity of concept and attribute selection is presented in [14]. In that work, the authors build a general framework for concept selection, in which the relevance criterion is represented as a closure operator. The relevant concepts are the fixpoints of the associated closure operator and they form a complete $\vee$-sublattice of the original concept lattice. Several constraints related to the cardinality of the extent and to the presence or absence of different attributes are also studied. The proposed method allows to compute the reduced lattice and extract minimal bases of attribute dependencies as well, with lower computational effort (polynomial delay complexity), since there is no need to compute all concepts or pseudo-intents beforehand.

A more recent approach [42] introduces a method for attribute selection in formal contexts based on the notion of attribute relevance: according to that definition, an attribute is relevant if and only if it is irreducible in the context. This allows to define a *relative relevance function* which captures both the order structure in the concept lattice and the distribution of objects.

The relative relevance function is not computationally feasible (it has a high complexity), so the problem is approximated using ideas from information theory, defining the *Shannon object information entropy of a formal context* and its *object information entropy*. It is experimentally tested and proved that the use of the entropy measures is appropriated in contexts with many attributes and reflects the relative relevance of attributes properly.

As a consequence, in a Big Data problem, this type of mechanisms based on the pre-definition of a relevance measure (probably based on information theory, due to its computational efficiency) which allows to restrict the computation of concepts to only the ones with a higher relevance, presents great potential of use, and can therefore help develop optimized methods to mine the knowledge in a large formal context.

9.3 Improved Management of Implications

The computation of closures is one of the core steps when reasoning from an implication system. It requires to apply repeatedly the implications in the system (usually the *canonical* basis) until getting to a fixpoint of the closure operator, which means it has exponential complexity.

The main reason that forces to apply several times all the implications in the system in order to compute a closure (causing the high computational complexity) is given by the application of the *transitivity* axiom in the logic [4]:

$$[\text{Tran}] \frac{A \to B, B \to C}{A \to C}$$

The transitivity rule somehow reflects the cut rule in other logical systems and, hence, is not suitable for automation. Even works defining equivalent axiom systems [5, 43] do not arrive at an appropriate way of handling its inherent complexity in an efficient manner.

As a consequence, there has been traditionally a necessity of finding efficient computational methods by modifying the axiom system (to avoid the computationally expensive transitivity [Tran]) and designing new inference and reasoning methods.

Another way to reduce the computational cost of computing closures is to define some modified implicational systems, equivalent to the Duquenne-Guigues basis [41], but with a simpler structure that could be exploited in the calculation of closures. Although Duquenne-Guigues basis has minimum cardinality, there are other parameters which can be used to define alternative minimality conditions (e.g. *directness* [17, 18]) which, in turn, might have better computational properties.

Simplification Logic

The investigation on defining axiom systems equivalent to Armstrong's rules, but removing the transitivity axiom, led to the idea of the Simplification Logic.

Simplification Logic SL_{FD} [27], and its fuzzy counterpart FASL [9], define a logic equivalent to Armstrong's Axioms that avoids the use of transitivity and is guided by the idea of simplifying the set of implications by removing redundancies, which can be defined in the following terms:

Definition 9.6 Let $\mathbb{K} = (G, M, I)$ be a formal context and $\Gamma = \{A \to B : A, B \subseteq M\}$ be an implicational system.

- An implication φ is *superfluous* in Γ if φ can be inferred from $\Gamma \smallsetminus \{\varphi\}$.
- $\varphi = X \to Y$ is *l-redundant* in Γ if there exists $\emptyset \neq Z \subseteq X$ such that φ can be inferred from $(\Gamma \smallsetminus \{\varphi\}) \cup \{(X \smallsetminus Z) \to Y\}$.
- $\varphi = X \to Y$ is *r-redundant* in Γ if there exists $\emptyset \neq Z \subseteq Y$ such that φ can be inferred from $(\Gamma \smallsetminus \{\varphi\}) \cup \{X \to (Y \smallsetminus Z)\}$.

We say that Γ has redundancy if it has a superfluous, *l*-redundant or *r*-redundant element.

SL_{FD} provides new substitution operators which allow the natural design of automated deduction methods and new substitution rules which can be used bottom-up and top-down to get equivalent sets of implications, but without redundancy.

Definition 9.7 The SL_{FD} system has one axiom:

$$[\text{Ax}] \frac{}{X \to Y} \quad \text{if } Y \subseteq X$$

and three inference rules (fragmentation, composition and substitution):

$$[\text{Frag}] \frac{X \to Y}{X \to Y'} \quad \text{if } Y' \subseteq Y$$

$$[\text{Comp}] \frac{X \to Y, U \to V}{X \cup U \to Y \cup V}$$

$$[\text{Subst}] \frac{X \to Y, U \to V}{U \smallsetminus Y \to V \smallsetminus Y} \quad \text{if } X \subseteq U, X \cap Y = \emptyset$$

The corresponding substitution operators associated to the [Subst] rule are given below.

Definition 9.8 Given an implication $X \to Y$:

- The *substitution operator* associated to $X \to Y$ is defined as:

$$\Phi_{X \to Y}(U \to V) = \begin{cases} U \smallsetminus Y \to V \smallsetminus Y & \text{if } X \subseteq U, X \cap Y = \emptyset \\ U \to V & \text{otherwise} \end{cases}$$

- The *right-substitution operator* associated to $X \to Y$ is defined as:

$$\Phi^{\mathrm{r}}_{X \to Y}(U \to V) = \begin{cases} U \to V \smallsetminus Y & \text{if } X \nsubseteq U, X \cap Y = \varnothing, X \subseteq U \cup V \\ U \to V & \text{otherwise} \end{cases}$$

The extension to the fuzzy setting was presented in [9]. In that case, the axiomatic system used the following inference rule (simplification)

$$[\mathtt{Sim}] \; \frac{X \to Y, U \to V}{X \cup (U \smallsetminus Y) \to V}$$

instead of [Subst].

The application of these operators generates a simpler implicational system, removing superfluous and *l*- and *r*-redundant implications.

The use of a *simplified* implication system can have a computational impact when making inference on *massive amounts of data*, since the number of operations to be performed decrease as redundancies are removed. For instance, this logic has proved to be useful for automated reasoning with implications [25, 28, 29, 59, 60]. However, the overall complexity of computing a closure remains exponential, in the worst-case scenario, as is in the case of the Duquenne-Guigues basis.

Direct Bases

One reason for the high complexity in computing closures is the need to make several applications of a whole implicational system until getting to a fixpoint, due to the *transitivity* axiom. It is reasonable to investigate the definition of implicational systems equivalent to a given one, such that closures can be computed in a single pass over the implication set [17, 18, 26, 69, 70].

This property for these systems is called *directness* [17, 18], and can be introduced formally as follows: suppose an implicational system Γ, over the set of attributes M, and define the operator $\pi_\Gamma : 2^M \to 2^M$ as $\pi_\Gamma(X) = X \cup \{b \in B | A \to B \in \Gamma \text{ for some } A \subseteq X\}$.

This function π_Γ is isotone and extensive and, therefore, for all $X \in 2^M$, the chain $X, \pi_\Gamma(X), \pi_\Gamma^2(X), \pi_\Gamma^3(X), \ldots$ reaches a fixpoint, and the closure of the set X coincides with this fixpoint. For specific implicational systems, π_Γ is idempotent, which means that the fixpoint is reached in the first iteration, i.e. with a single traversal of the implicational system [26].

Thus, inference with this type of bases can be done in a time complexity which is linear with respect to the number of the implications in the basis.

If, to the condition of a system Γ being direct, one adds that both its cardinality $|\Gamma|$ (number of implications in it) and its size [53] ($\|\Gamma\| = \sum_{A \to B \in \Gamma}(|A| + |B|)$) are minimum among all equivalent systems, an important reduction of the complexity of computing closures may be achieved.

Definition 9.9 A direct implicational system Γ is said to be a *direct-optimal* basis if, for any direct implicational system Γ', equivalent to Γ, one has $\|\Gamma\| \leq \|\Gamma'\|$.

It is proved in [17] that, for any implicational system Γ, there exists a unique direct-optimal basis $\Gamma_{do} \equiv \Gamma$.

Direct-optimal bases combine the directness and optimality properties. On the one hand, directness ensures that the computation of the closure may be done in just one traversal of the implication set; on the other hand, due to its minimal size provided by the optimality, the number of visited implications is reduced to the minimum. Due to these features, it is desirable to design methods to transform an arbitrary set of implications into its equivalent direct-optimal basis. Thus, the problem of building a direct-optimal basis is one of the outstanding problems in FCA.

Several schemes have been proposed to compute the direct-optimal basis [17, 70], by studying the possibility of using unitary implications. In those cases, although the output implicational system is of higher cardinality (since all *conclusions* are unitary), its computation is more efficient.

In the general (*n*-ary) case, in [69], a simple algorithm to compute such a direct-optimal basis is presented, based on the sequential application of the rules of the previous Simplification Logic SL_{FD}. A new rule can be obtained which improves the *overlap* rule Ovl from [18]

$$[Ovl] \frac{A \rightarrow B, C \rightarrow D}{A \cup (C \smallsetminus B) \rightarrow D} \quad \text{if } B \cap C \neq \varnothing$$

the new rule, called *strong simplification*, can be formally derived from SL_{FD} and is used iteratively in conjunction with the *(right-)substitution operator*, leading to simpler implications.

$$[sSimp] \frac{A \rightarrow B, C \rightarrow D}{A \cup (C \smallsetminus B) \rightarrow D \smallsetminus (A \cup B)} \quad \text{if } B \cap C \neq \varnothing$$

Its associated *strong-simplification operator* is the core of the method to compute the unique direct-optimal basis.

The complexity of the calculation of such direct-optimal basis is still exponential in the number of implications in the original implication system in the worst-case. One may argue that, in this case, there may be no gain in a practical situation. However, in practice [69], it is asymptotically faster (i.e., when the number of implications or attributes is increased) with respect to the previous most efficient algorithm [70].

In addition, note that in this case the computation of closures is reduced to a single pass over a (larger in size) implicational system. Thus, when reasoning in a Big Data context, the use of direct-optimal bases may be more efficient. Since the computationally expensive construction of such a basis can be done *offline*, this overhead does not impact the end user. Then, the main benefit is that the computation of closures using this basis can be done in the time constraints currently imposed in Big Data settings.

Ordered-Direct Bases

Given a set of indexed implications $\Gamma = \{A_i \to B_i\}$, $1 \le i \le n$, Adaricheva et al. [1] introduced an alternative approach to directness, named *ordered-directness*, in terms of the *ordered iteration operator*, which is defined by $\rho_\Gamma = \pi_{\Gamma_n} \circ \ldots \pi_{\Gamma_1}$, where $\Gamma_i = \{A_i \to B_i\}$ for all i.

Analogously to the π_Γ operator, ρ_Γ has the same computational cost and it is isotone and extensive. In addition, one can get that, for all $X \subseteq M$, it holds $\pi_\Gamma(X) \subseteq \rho_\Gamma(X)$, what can lead to a faster convergence to a fixpoint of the chain $X, \rho_\Gamma(X), \rho_\Gamma^2(X), \ldots$

Definition 9.10 [1] An implicational system Γ is said to be *ordered direct* if ρ_Γ is idempotent, that is, if $\rho_\Gamma(X)$ coincides with the closure of X with respect to Γ, for all $X \subseteq M$.

In this case, it must be noted that the directness of an implicational system implies its ordered-directness, but the converse is not always true [1]. In the same work, the authors propose a new kind of bases, called the D-bases, by using the ρ_Γ operator.

Definition 9.11 [1] Let Γ be a reduced implicational system (that is, $A \cap B = \varnothing$ for all $A \to B \in \Gamma$), and let us denote by A_Γ^+ the closure of the set A with respect to Γ. The D-basis for Γ is the pair $\langle \Gamma_a, \Gamma_n \rangle$, where:

- $\Gamma_a = \{x \to y \mid y \in M, x \in \{y\}_\Gamma^+, x \neq y\}$
- $\Gamma_n = \{X \to x \mid X \subseteq M, x \in M, X \text{ is a minimal proper cover of } x\}$

Here, the notion of a set $X \subseteq M$ being a *proper cover* of $y \in M$ with respect to the implicational system Γ means that $y \in X_\Gamma^+ \setminus \left(\bigcup_{x \in X} \{x\}_\Gamma^+ \right)$, i.e., the closure of X contains y, but no single element $x \in X$ has y in its closure. This implies that, evidently, $y \notin X$.

A *proper cover* $X \subseteq M$ of $y \in M$ is said to be *minimal* if, for all other proper cover Z of y, with $Z \subseteq \bigcup_{x \in X} \{x\}_\Gamma^+$, we have $Z \subseteq X$.

Note that implications in Γ_a are atomic, that is, both premise and conclusion are singletons (atoms of M). Implications in Γ_n have unitary conclusion but n-ary premise.

The next result states that D-bases are actually a subclass of *ordered-direct* bases.

Theorem 9.2 ([1]) *Let Γ be an implicational system, with $\langle \Gamma_a, \Gamma_n \rangle$ its associated D-basis and let $\Gamma_D = \Gamma_a \cup \Gamma_n$, ordered in such a way that, in the ordered iteration operator ρ_{Γ_D}, atomic implications from Γ_a are checked before those of Γ_n. Then Γ_D is an ordered-direct basis equivalent to Γ.*

In addition, if Γ_{udo} is the unitary (relative to conclusions) direct-optimal basis equivalent to Γ, then $\Gamma_D \subseteq \Gamma_{udo}$.

D-bases belong to the family of bases whose implications have unitary conclusions. In the mentioned work [1], a method is proposed to extract Γ_D from any direct unit basis Γ_{udo} in polynomial time with respect to the size of Γ_{udo}, and taking only linear time (of the cardinality of the produced Γ_D) to put it into the order assumed in the previous theorem (atomic implications before n-ary ones).

A further improvement was made in [68], where the idea of aggregating a D-basis is presented. A D-basis is *aggregated* if its premises are pairwise disjoint. In that work, the uniqueness of aggregated D-bases is proved. By using the simplification logic, the authors propose the fastD-basis algorithm, whose input is an implication system, not necessarily a direct unit basis, and returns the unique aggregated D-basis equivalent to it. Also, the authors proved empirically a higher performance with respect to the previous technique [1].

Running the D-basis in one iteration is more efficient than running a shorter, but unordered, canonical basis, such as Duquenne-Guigues. There are examples demonstrating that the canonical basis cannot always be ordered [1].

As stated above, since the computation of an implication basis can be made *offline* in practical Big Data scenarios, the main goal in this research line is to obtain the most adequate presentation of an implication system which allows to perform the computation of closures fast enough to be used in a real-world application. A D-basis presents a promising step towards the development of a knowledge engine completely automated and applicable in real-world situations.

9.4 Minimal Generators to Represent Knowledge

A compact representation of closed sets in a closure system can have an important impact on the efficient evaluation and construction of implicational bases.

In this sense, minimal generators[8] (or *mingens*) constitute a key part of the closure structure, since they represent minimal sets in the underlying equivalence relation over subsets of the attribute set M.

Definition 9.12 Given a closed set $X \subseteq M$ and a set of implications Γ over M, a subset $Y \subseteq M$ is called a *generator* of X if $X = Y_\Gamma^+$, that is, if the closure of Y with respect to the implicational system Γ is equal to X.

Note that any other subset of X containing its generator Y is also a generator of X. As the set of attributes M is finite, the set of its generators can be characterized by looking for those with the minimality condition.

Definition 9.13 Let M be a finite set of attributes and Γ an implicational system on M. $X \subseteq M$ is called a *minimal generator* (*mingen*) if, for all proper subsets $Y \subsetneq X$, it holds that $Y_\Gamma^+ \subsetneq X_\Gamma^+$.

For a given closed set $C \subseteq M$, let us denote as $mg(C)$ the set its minimal generators, therefore $mg : 2^M \to 2^{2^M}$.

Minimal generators [8] were introduced in various fields under various names: minimal keys in the database field [54], irreducible gaps [41], minimal blockers [64] and 0-free itemsets [21].

The importance of minimal generators is that they store a compact representation of the knowledge stored in a context. Minimal generators favor the principle of *minimum description length*, that is, the best hypothesis for a given dataset is the one leading to the best compression of the data [40].

A remarkable contribution on the relevance of *mingens* is that they can be used to build the *iceberg* lattice [72] using the TITANIC algorithm, a smart procedure that is able to take advantage of minimal generators (or *key sets* as they are called in that work) to generate the concept lattice of minimal generators.

This kind of lattices has been successfully used in several applications, such as the analysis of large databases, and extracting implications and mining association rules [32]. This is specially useful when dealing with *large* datasets, since the algorithm is able to retrieve *mingens* with high support.

Also, from this set of minimal generators, we can rebuild all the information that may be inferred from the context, in the form of an implicational basis. Due to this compact representation, the derived bases may allow for a better performance of the reasoning methods.

In [35, 61], some methods for computing the set of *mingens* have been proposed. In parallel to these methods, in [57, 58] minimal generators are used to compute implication bases (on contexts with positive and negative attributes) whose premises are minimal generators.

In all the previous works, the context was considered as the input of the problem, that is, the set of minimal generators (and of closed sets) was inferred directly from the dataset. A more logic-oriented approach was used in [30], where the presented method allowed to derive the set of *mingens* from an implication basis, using the previously mentioned Simplification Logic, instead of using the context as starting point. This complete and precise specification allows for faster managing the semantics of the information contained in the dataset.

Another technique was developed in [28], where the MinGen algorithm was designed to build a search tree space (of the implications) that can be traversed (using inference rules) to find all the minimal generators. This shape of the search space limits its execution for medium-sized problems, because of the overwhelming requirements of the sequential MinGen algorithm.

In order to get this algorithm working in medium-sized problems, in [16] the authors present an efficient reduction of the search space technique to improve the performance of the enumeration of *mingens*. The new method was designed to fit the Map-Reduce architecture, and thus parallel computation makes it possible to deal with large datasets. As the authors state in their work, "the empirical study proves the very significant improvement achieved w.r.t the original sequential version. The parallel methods to compute minimal generators can make really usable these methods in practical applications."

As commented before, *minimal generators* can be used to rebuild an implication basis. To this end, the notion of labelled set of items is introduced:

Definition 9.14 [28] A *labelled set of items* (LSI) is a collection $\Phi = \{\langle A_i, B_i \rangle\}$, where $A_i \subseteq M$ and $B_i \subseteq 2^{A_i}$ such that if $X, Y \in B_i$ with $X \subseteq Y$ then $X = Y$.

Particularly, for an arbitrary set of implications Γ, we can consider the LSI $\Phi = \{\langle C, \mathrm{mg}(C) \rangle \mid C$ is a closed set under $\Gamma\}$ of special interest, since it can be used [29] as input to a procedure aimed to extract a so-called *left-minimal* basis, defined as follows.

Definition 9.15 An implicational system Γ is a *left-minimal* basis if there is no $A \to B \in \Gamma$ and $A' \subsetneq A$ such that $\Gamma \smallsetminus \{A \to B\} \cup \{A' \to B\}$ is equivalent to Γ. In addition, Γ is *direct* if, for all $A \to B \in \Gamma$, one has that $A \cup B$ is closed with respect to Γ.

In other words, a *left-minimal direct* basis is a set of implications where premises are minimal generators and the corresponding conclusions are their associated closed sets. Such a *left-minimal direct* basis has the minimality property of canonical bases, and the characteristics of the implications described before: minimal information in the left hand side and a fast computation of closures.

In order to compute the basis, the following result, given in [29], states that two aggregation operators can be used iteratively.

Theorem 9.3 *Let* $T = \{(M_i, A_i \to B_i) \ldots\}$ *be a set of pairs with minimal generators and implications obtained from minimal generators and closed sets. The exhaustive application of the two following aggregation rules:*

- *If* $A \subseteq C$, *then* $\{A \to B, C \to D\} \equiv \{A \to B, B \cup C \to D \smallsetminus B\}$.
- *If* $A \subseteq C \subseteq A \cup B$, *then* $\{A \to B, C \to D\} \equiv \{A \to B \cup D\}$.

produces a left-minimal direct basis.

Using this result, Cordero et al. propose in [29] an extension of the algorithm designed to compute the Duquenne-Guigues basis [12, 41], in order to obtain the left-minimal basis. This algorithm runs in polynomial time on the length of the LSI used as input. The first step is to compute an implication set from the LSI of the minimal generators (as the one described in the previous theorem) and, then, apply the aggregation rules to manipulate the implications and obtain a left-minimal basis.

9.5 Probably Approximately Correct Implication Bases

Current algorithms to find implication bases have an enormous overhead since they have to find *all* closed sets as a necessary step in their execution. This has led to another research line in which the exactness of the implication basis to be found is not considered as fundamental, that is, it is allowed to have non-exact or non-complete but representative and informative bases.

Besides this difficulty in computing implication bases, there is another more practical reason to relax the constraint of having exact implications: real-world datasets are generally noisy, containing errors and inaccuracies, therefore computing exact implication bases from such datasets may be useless or even a nonsense. In this case, a rather pragmatic approach is to consider implications as strong association rules and then use highly optimized association rule algorithms [2], but the number of resulting implications increases even more.

Some works [6, 19, 20] have studied approximations of the exact implication basis, taking into account that such bases should have a controllable error, otherwise they become useless. Their proposal is, instead of calculating large exact bases, to compute *approximately correct bases* that could capture most of the implicational theory of a given dataset (or at least the most essential parts) but are easier to compute.

It can be said [20] that a set Ω of implications is an *approximately correct basis* of the formal context $\mathbb{K}$ if most closed sets of Ω are closed in $\mathbb{K}$ and vice versa. The formal idea behind this intuition on *approximate* bases is to define a measure of *proximity* between sets of implications in terms of their closed sets. Thus, the key to build approximate bases is to define a *distance* between closure operators.

In [6], one can find initial results on approximate bases and some experimental evaluations, and a set of implications Ω is defined as an *approximation* of another set Γ if the closure operators of both coincide on most subsets of the attribute set M. This measure can be defined in terms of the cardinality of $\{S \subseteq M : \Omega(S) \neq \Gamma(S)\}$, being $\Omega(S)$ and $\Gamma(S)$ the closures of S with respect to both sets of implications. This can be understood from the application point of view, since that definition ensures that, in most cases, operating with Γ and with Ω will provide the same closures.

This work has been extended to the idea of building *probably approximately correct bases* (PAC bases) in [20], which are approximately correct with high probability. This notion gains strength, since PAC bases can be computed in polynomial time. In this new approach, the concept of *approximation* is slightly different from the one in [6]. Ω is an approximation of Γ if and only if the number of closed sets in which Ω and Γ differ is small. More precisely, they define an *approximately correct basis* as follows.

Definition 9.16 [20] Let M be a finite set and let $\mathbb{K} = (G, M, I)$ be a formal context. A set of implications Ω is a *approximately correct basis* for $\mathbb{K}$ with accuracy $\varepsilon > 0$ if

$$d(\Omega, \mathbb{K}) := \frac{|\mathrm{cl}(\Omega) \triangle \mathrm{cl}(\mathbb{K})|}{2^{|M|}} < \varepsilon$$

where $\mathrm{cl}(\Omega)$ and $\mathrm{cl}(\mathbb{K})$ are the sets of closed sets of Ω and $\mathbb{K}$, respectively, and $S_1 \triangle S_2$ represents the set symmetric difference of S_1 and S_2. $d(\Omega, \mathbb{K})$ is called the *Horn distance* between Ω and $\mathbb{K}$.

And, from this definition, they build the idea of a PAC basis as follows.

Definition 9.17 [20] Let M be a finite set and let $\mathbb{K} = (G, M, I)$ be a formal context, let $\mathrm{Imp}(M)$ be the set of all possible implications between the elements of M, and let $\mathscr{O} = (\mathscr{W}, \mathscr{E}, Pr)$ be a probability space. A random variable $\Omega : \mathscr{O} \to 2^{\mathrm{Imp}(M)}$ is called a *probably approximately correct basis* (PAC basis) of $\mathbb{K}$ with accuracy $\varepsilon > 0$ and confidence $\delta > 0$ if $Pr(d(\Omega, \mathbb{K}) > \varepsilon) < \delta$.

The computational efficiency of this approach comes from the fact that it is a modified version of the Horn algorithm [3], making use of *membership* and *equivalence oracles* (playing the role of *domain experts*), which allows to compute PAC bases in polynomial time in size of M, the output Ω, as well as $\frac{1}{\varepsilon}$ and $\frac{1}{\delta}$ (ε and δ are inputs to the algorithm), provided that the invocations of the oracles are counted as single steps [20]. In this work, the authors test the usability of PAC bases in real-world situations, comparing several measures of the practical quality of the approximation: the Horn distance between the canonical basis and the approximating bases, and the usual precision and recall measures, defined as follows:

Definition 9.18 [20] Let M be a finite set, let $\mathbb{K} = (G, M, I)$ be a formal context and let Γ be the canonical basis of $\mathbb{K}$. The *precision* and *recall* of a basis Ω, are defined as:

$$\mathrm{prec}(\mathbb{K}, \Omega) = \frac{|\{(A \to B) \in \Omega : \Gamma \models A \to B\}|}{|\Omega|}$$

$$\mathrm{recall}(\mathbb{K}, \Omega) = \frac{|\{(A \to B) \in \Gamma : \Omega \models A \to B\}|}{|\Gamma|}$$

From this definition, *precision* measures the fraction of valid implications in the approximating basis, and *recall* measures the fraction of valid implications in the canonical basis Γ that follow semantically from the approximating basis Ω.

It has been found [20] that increasing the value of ε in the algorithm always leads to a considerable increase in the Horn distance, meaning that the PAC basis deviates more and more from the canonical basis. In addition, the theoretical upper bound ε for the Horn distance between Ω and the canonical basis is never realized in their experiments, meaning that the obtained PAC basis is indeed much closer to the canonical basis than what the algorithm is theoretically designed to obtain. Lastly, for small values of ε (the choice of δ seems to have little impact in the experimental results), both precision and recall are very high, i.e., close to one, what means that the algorithm is able to retrieve most of the canonical basis, and that most of the implications of Ω are valid.

The important result that PAC bases can be computed in output-polynomial time opens the way to decrease the long running times of the algorithms to compute *exact* implication bases. Thus, this is a promising line in the integration of FCA techniques into Big Data situations, as the applicability of the former to larger datasets become feasible.

9.6 Summary and Possible Future Trends

As stated above, there are important reasons which suggest the convenience of further developing logical methods for FCA if we are targeting big datasets: the main issues here are the explainability and interpretability, which are inherent to logic-based methods but are missing in the usual machine learning tools for Big Data.

In this work, we have surveyed some trends in FCA that could be of potential application in Big Data settings, many of them are focused on alleviating the high computational cost of current methods to build the concept lattice, to find a basis of implications and to reason with it.

First, we have considered some techniques to remove redundant information in a formal context, such as clarification and reduction, or the search for reducts through the computation of the discernibility matrix. The aim of these techniques is to build a simpler formal context, with exactly the same closure space as the original one, where computations are less expensive since the number of objects or attributes is reduced. The techniques based on the discernibility matrix are, in the best case, equivalent to clarification and reduction, but with a complexity that is exponential in the worst case. Thus, it depends on the particular application which of the two approaches is more suitable.

Other techniques, based on matrix factorization methods, study the problem of simplifying the structure of the formal context, preserving most of the information, but not all.

The selection of relevant attributes (and concepts) is another open problem, since the measure of relevance is application-dependent. In this sense, the Titanic algorithm is used to build *iceberg* lattices which consists only of concepts with support above a predefined threshold. Since it is more computationally efficient than computing the whole concept lattice, it is potentially applicable to real-world Big Data problems.

With respect to the ability to reason and make inference using implication systems, there is a blocking issue which makes the use of logic tools difficult in Big Data. The presentation of an implication system, together with the axiomatization of the logic (more precisely, the transitivity axiom), makes the computation of closures in Big Data an unsolved issue. We have collected some mechanisms which could potentially help reduce the computational overhead of using logic in Big Data. First, the Simplification Logic can be used to remove atomic redundancies in implication systems. The axiomatization of the Simplification Logic removes the need of the transitivity axiom, thus providing a mechanism to get simpler implication systems which could be traversed in a more efficient manner.

In this direction, other types of implication bases are defined. Particularly, *direct*, *ordered-direct* and D- bases are implication bases, equivalent to the canonical basis, which only require a single pass over the implication set to compute a closure. Since the computation of one of these implication bases can be made *offline* in practical Big Data scenarios, an adequate presentation of the implication system which allows to perform the calculation of closures with enough speed could be used in practical real-world situations.

Regarding the presentation of the knowledge, a compact representation of the closure system of a Big Data formal context can be useful. In this line, minimal generators are at the core of the closure space, and they have been used (e.g. in the Titanic algorithm) to compute simplified versions of the concept lattice, and to rebuild all information that can be inferred from the context.

The last line that has been explored in this work is the computation of *approximately correct* implication bases. Since in practical situations it may be neither advisable nor useful to compute a complete basis of *exact* implications, due either to the prohibiting computational cost or to the noise that might be present in data, it is more pragmatic to capture just *most* of the implication theory. This is the idea behind *approximately correct* bases, which tend to cover most of the closure space of the canonical basis, with valid implications.

This idea has been extended to present *probably approximately correct* bases, which are approximately correct *with high probability*. In this approach, the main strength is that, by using *oracles* in the role of *domain experts*, this kind of bases can be computed in polynomial time. Also, experimentally, PAC bases have presented a good practical quality, in terms of its similarity to the canonical basis.

Certainly, the computational complexity of many problems easily exceeds the tractability threshold (i.e. makes FCA unusable in real-world Big Data problems), hence it makes no sense to expect a complete logic-driven FCA approach; but, what about trying to hybridize techniques? One could wonder to develop, firstly, formal and logic-driven methods up to certain level and, then, applying machine learning techniques, making use of alternative data structures and parallelization techniques. It could be worth to further push the research line of using neural networks to implement the closure operators directly from the context.

Continuing with this line, further development of reduction methods for the contexts is essential, and here, one could consider, for instance, approaching the dimensionality reduction via principal components analysis (PCA) in terms of fuzzy computing instead of numerical PCA. Since PCA is a space decomposition technique, it may be interesting to study how it could be applied to decompose a closure space or a formal context to reduce the computational overhead of present methods. In this case, it is worth remarking that the precise computation of a *fuzzy* concept lattice is more complex than in the crisp case but, following the analogy with fuzzy control, which has proven to be successful to handle very complex systems, it could be interesting to approach FCA by computing in terms of linguistic variables.

Acknowledgements This work has been partially supported by the Spanish Ministry of Science, Innovation, and Universities (MCIU), the State Agency of Research (AEI), the Junta de Andalucía (JA), the Universidad de Málaga (UMA), and the European Social Fund (FEDER) through the research projects with reference PGC2018-095869-B-I00 (MCIU/AEI/FEDER, UE) and UMA2018-FEDERJA-001 (JA/UMA/FEDER, UE)

References

1. Adaricheva, K.V., Nation, J.B., Rand, R.: Ordered direct implicational basis of a finite closure system. Discrete Applied Mathematics **161**(6), 707–723 (2013)
2. Agrawal, R., Srikant, R.: Fast algorithms for mining association rules in large databases. In: VLDB, pp. 487–499. Morgan Kaufmann (1994)
3. Angluin, D.: Queries and concept learning. Mach. Learn. **2**(4), 319–342 (1987)
4. Armstrong, W.W.: Dependency structures of data base relationships. In: J.L. Rosenfeld (ed.) Information Processing, Proceedings of the 6th IFIP Congress 1974, Stockholm, Sweden, August 5–10, 1974., pp. 580–583. North-Holland (1974)
5. Atzeni, P., Antonellis, V.D.: Relational Database Theory. Benjamin/Cummings (1993)
6. Babin, M.A.: Models, methods, and programs for generating relationships from a lattice of closed sets. Ph.D. thesis, Higher School of Economics, Moscow (2012)
7. Babin, M.A., Kuznetsov, S.O.: Computing premises of a minimal cover of functional dependencies is intractable. Discrete Applied Mathematics **161**(6), 742–749 (2013)
8. Bastide, Y., Pasquier, N., Taouil, R., Stumme, G., Lakhal, L.: Mining minimal non-redundant association rules using frequent closed itemsets. In: Computational Logic, *Lecture Notes in Computer Science*, vol. 1861, pp. 972–986. Springer (2000)
9. Belohlavek, R., Cordero, P., Enciso, M., Mora, Á., Vychodil, V.: Automated prover for attribute dependencies in data with grades. International Journal of Approximate Reasoning **70**, 51–67 (2016)
10. Belohlávek, R., Macko, J.: Selecting important concepts using weights. In: ICFCA, *Lecture Notes in Computer Science*, vol. 6628, pp. 65–80. Springer (2011)
11. Belohlávek, R., Outrata, J., Trnecka, M.: Factorizing boolean matrices using formal concepts and iterative usage of essential entries. Inf. Sci. **489**, 37–49 (2019)
12. Belohlávek, R., Vychodil, V.: Formal concept analysis with constraints by closure operators. In: ICCS, *Lecture Notes in Computer Science*, vol. 4068, pp. 131–143. Springer (2006)
13. Belohlávek, R., Vychodil, V.: Discovery of optimal factors in binary data via a novel method of matrix decomposition. J. Comput. Syst. Sci. **76**(1), 3–20 (2010)
14. Belohlávek, R., Vychodil, V.: Closure-based constraints in formal concept analysis. Discrete Applied Mathematics **161**(13–14), 1894–1911 (2013)
15. Benítez-Caballero, M.J., Medina, J., Ramírez-Poussa, E.: Attribute reduction in rough set theory and formal concept analysis. In: IJCRS (2), *Lecture Notes in Computer Science*, vol. 10314, pp. 513–525. Springer (2017)
16. Benito-Picazo, F., Cordero, P., Enciso, M., Mora, A.: Minimal generators, an affordable approach by means of massive computation. The Journal of Supercomputing **75**(3), 1350–1367 (2019)
17. Bertet, K., Monjardet, B.: The multiple facets of the canonical direct unit implicational basis. Theoretical Computer Science **411**(22–24), 2155–2166 (2010)
18. Bertet, K., Nebut, M.: Efficient algorithms on the moore family associated to an implicational system. Discrete Mathematics and Theoretical Computer Science **6**(315–338), 107 (2004)
19. Borchmann, D.: Learning terminological knowledge with high confidence from erroneous data. Ph.D. thesis, Saechsische Landesbibliothek-Staats-und Universitaetsbibliothek Dresden (2014)
20. Borchmann, D., Hanika, T., Obiedkov, S.: On the usability of probably approximately correct implication bases. In: ICFCA, *Lecture Notes in Computer Science*, vol. 10308, pp. 72–88. Springer (2017)
21. Boulicaut, J., Bykowski, A., Rigotti, C.: Free-sets: A condensed representation of boolean data for the approximation of frequency queries. Data Min. Knowl. Discov. **7**(1), 5–22 (2003)

22. Caruana, R., Lou, Y., Gehrke, J., Koch, P., Sturm, M., Elhadad, N.: Intelligible models for healthcare: Predicting pneumonia risk and hospital 30-day readmission. In: Proceedings of the 21th ACM SIGKDD International Conference on Knowledge Discovery and Data Mining, pp. 1721–1730. ACM (2015)

23. Cheung, K.S.K., Vogel, D.R.: Complexity reduction in lattice-based information retrieval. Inf. Retr. 8(2), 285–299 (2005)

24. Codocedo, V., Taramasco, C., Astudillo, H.: Cheating to achieve formal concept analysis over a large formal context. In: CLA, *CEUR Workshop Proceedings*, vol. 959, pp. 349–362. CEUR-WS.org (2011)

25. Cordero, P., Enciso, M., López, D., Mora, A.: A conversational recommender system for diagnosis using fuzzy rules. Expert Systems with Applications p. 113449 (2020)

26. Cordero, P., Enciso, M., Mora, A.: Directness in fuzzy formal concept analysis. In: Proc. of the 17th Intl Conf on Information Processing and Management of Uncertainty in Knowledge-Bases IPMU'18, *Communications in Computer and Information Science*, vol. 853, pp. 585–595 (2018)

27. Cordero, P., Enciso, M., Mora, A., de Guzmán, I.P.: SL_{FD} logic: Elimination of data redundancy in knowledge representation. In: Ibero-American Conference on Artificial Intelligence, pp. 141–150. Springer (2002)

28. Cordero, P., Enciso, M., Mora, A., Ojeda-Aciego, M.: Computing minimal generators from implications: a logic-guided approach. In: CLA, vol. 2012, pp. 187–198. Citeseer (2012)

29. Cordero, P., Enciso, M., Mora, A., Ojeda-Aciego, M.: Computing left-minimal direct basis of implications. In: CLA, vol. 2013, pp. 293–298 (2013)

30. Cordero, P., Enciso, M., Mora, A., Ojeda-Aciego, M., Rossi, C.: Knowledge discovery in social networks by using a logic-based treatment of implications. Knowl.-Based Syst. 87, 16–25 (2015)

31. Dias, S.M., Vieira, N.J.: Concept lattices reduction: Definition, analysis and classification. Expert Syst. Appl. 42(20), 7084–7097 (2015)

32. Dias, S.M., Z'arate, L.E., Vieira, N.J.: Using iceberg concept lattices and implications rules to extract knowledge from ANN. Intelligent Automation & Soft Computing 19(3), 361–372 (2013)

33. Distel, F.: Hardness of enumerating pseudo-intents in the lectic order. In: International Conference on Formal Concept Analysis, pp. 124–137. Springer (2010)

34. Distel, F., Sertkaya, B.: On the complexity of enumerating pseudo-intents. Discrete Applied Mathematics 159(6), 450–466 (2011)

35. Dong, G., Jiang, C., Pei, J., Li, J., Wong, L.: Mining succinct systems of minimal generators of formal concepts. In: DASFAA, *Lecture Notes in Computer Science*, vol. 3453, pp. 175–187. Springer (2005)

36. Dumbill, E.: What is big data? an introduction to the big data landscape. Strata 2012: Making Data Work (2012)

37. Gajdos, P., Moravec, P., Snásel, V.: Concept lattice generation by singular value decomposition. In: CLA, *CEUR Workshop Proceedings*, vol. 110. CEUR-WS.org (2004)

38. Ganter, B., Wille, R.: Formal Concept Analysis - Mathematical Foundations. Springer (1999)

39. Geerts, F., Missier, P., Paton, N.W.: Editorial: Special issue on improving the veracity and value of big data. J. Data and Information Quality 9(3), 13:1–13:2 (2018)

40. Grünwald, P.D., Grunwald, A.: The minimum description length principle. MIT press (2007)

41. Guigues, J.L., Duquenne, V.: Familles minimales d'implications informatives résultant d'un tableau de données binaires. Mathématiques et Sciences humaines 95, 5–18 (1986)

42. Hanika, T., Koyda, M., Stumme, G.: Relevant attributes in formal contexts. In: ICCS, *Lecture Notes in Computer Science*, vol. 11530, pp. 102–116. Springer (2019)

43. Ibaraki, T., Kogan, A., Makino, K.: Inferring minimal functional dependencies in horn and q-horn theories. Ann. Math. Artif. Intell. 38(4), 233–255 (2003)

44. Kauer, M., Krupka, M.: Removing an incidence from a formal context. In: CLA, *CEUR Workshop Proceedings*, vol. 1252, pp. 195–206. CEUR-WS.org (2014)

45. Khardon, R.: Translating between horn representations and their characteristic models. Journal of Artificial Intelligence Research **3**, 349–372 (1995)

46. Kim, B.: Interactive and interpretable machine learning models for human machine collaboration. Ph.D. thesis, Massachusetts Institute of Technology (2015)

47. Kim, B., Khanna, R., Koyejo, O.O.: Examples are not enough, learn to criticize! criticism for interpretability. In: Advances in Neural Information Processing Systems, pp. 2280–2288 (2016)

48. Konecny, J.: On attribute reduction in concept lattices: Methods based on discernibility matrix are outperformed by basic clarification and reduction. Inf. Sci. **415**, 199–212 (2017)

49. Konecny, J., Krajca, P.: On attribute reduction in concept lattices: The polynomial time discernibility matrix-based method becomes the cr-method. Inf. Sci. **491**, 48–62 (2019)

50. Kumar, C.A., Dias, S.M., Vieira, N.J.: Knowledge reduction in formal contexts using non-negative matrix factorization. Mathematics and Computers in Simulation **109**, 46–63 (2015)

51. Kuznetsov, S.O.: On the intractability of computing the duquenne-guigues basis. J. UCS **10**(8), 927–933 (2004)

52. Kuznetsov, S.O., Makhalova, T.P.: On interestingness measures of formal concepts. Inf. Sci. **442–443**, 202–219 (2018)

53. Lorenzo, E.R., Cordero, P., Enciso, M., Bonilla, A.M.: Canonical dichotomous direct bases. Inf. Sci. **376**, 39–53 (2017)

54. Maier, D.: The Theory of Relational Databases. Computer Science Press (1983)

55. Medina, J.: Relating attribute reduction in formal, object-oriented and property-oriented concept lattices. Computers & Mathematics with Applications **64**(6), 1992–2002 (2012)

56. Miller, T.: Explanation in artificial intelligence: Insights from the social sciences. Artificial Intelligence **267**, 1–38 (2019)

57. Missaoui, R., Nourine, L., Renaud, Y.: An inference system for exhaustive generation of mixed and purely negative implications from purely positive ones. In: M. Kryszkiewicz, S.A. Obiedkov (eds.) Proceedings of the 7th International Conference on Concept Lattices and Their Applications, Sevilla, Spain, October 19–21, 2010, *CEUR Workshop Proceedings*, vol. 672, pp. 271–282. CEUR-WS.org (2010)

58. Missaoui, R., Nourine, L., Renaud, Y.: Computing implications with negation from a formal context. Fundam. Inform. **115**(4), 357–375 (2012)

59. Mora, A., Cordero, P., Enciso, M., Fortes, I., Aguilera, G.: Closure via functional dependence simplification. International Journal of Computer Mathematics **89**(4), 510–526 (2012)

60. Mora, A., Enciso, M., Cordero, P., de Guzmán, I.P.: An efficient preprocessing transformation for functional dependencies sets based on the substitution paradigm. In: Conference on Technology Transfer, pp. 136–146. Springer (2003)

61. Nishio, N., Mutoh, A., Inuzuka, N.: On computing minimal generators in multi-relational data mining with respect to θ-subsumption. In: ILP (Late Breaking Papers), *CEUR Workshop Proceedings*, vol. 975, pp. 50–55. CEUR-WS.org (2012)

62. Osicka, P., Trnecka, M.: Boolean matrix decomposition by formal concept sampling. In: CIKM, pp. 2243–2246. ACM (2017)

63. Pawlak, Z.: Rough sets. Int. J. Parallel Program. **11**(5), 341–356 (1982)

64. Pfaltz, J.L.: Incremental transformation of lattices: A key to effective knowledge discovery. In: ICGT, *Lecture Notes in Computer Science*, vol. 2505, pp. 351–362. Springer (2002)

65. Polanyi, M.: The Tacit Dimension. Doubleday, Garden City, NY (1966)

66. Portugal, I., Alencar, P.S.C., Cowan, D.D.: The use of machine learning algorithms in recommender systems: A systematic review. Expert Syst. Appl. **97**, 205–227 (2018)

67. Qi, J.J.: Attribute reduction in formal contexts based on a new discernibility matrix. Journal of applied mathematics and computing **30**(1–2), 305–314 (2009)

68. Rodríguez-Lorenzo, E., Adaricheva, K.V., Cordero, P., Enciso, M., Mora, A.: Formation of the d-basis from implicational systems using simplification logic. Int. J. General Systems **46**(5), 547–568 (2017)
69. Rodríguez-Lorenzo, E., Bertet, K., Cordero, P., Enciso, M., Mora, Á.: Direct-optimal basis computation by means of the fusion of simplification rules. Discrete Applied Mathematics **249**, 106–119 (2018)
70. Rodríguez-Lorenzo, E., Bertet, K., Cordero, P., Enciso, M., Mora, A., Ojeda-Aciego, M.: From implicational systems to direct-optimal bases: A logic-based approach. Applied Mathematics & Information Sciences. L **2**, 305–317 (2015)
71. Rudolph, S.: Using FCA for encoding closure operators into neural networks. Lecture Notes in Computer Science **4604**, 321–332 (2007)
72. Stumme, G., Taouil, R., Bastide, Y., Pasquier, N., Lakhal, L.: Computing iceberg concept lattices with Titanic. Data Knowl. Eng. **42**(2), 189–222 (2002)
73. Valverde-Albacete, F.J., Peláez-Moreno, C., Cordero, P., Ojeda-Aciego, M.: Formal equivalence analysis. In: 2019 Conference of the International Fuzzy Systems Association and the European Society for Fuzzy Logic and Technology (EUSFLAT 2019). Atlantis Press (2019)
74. Wei, W., Wu, X., Liang, J., Cui, J., Sun, Y.: Discernibility matrix based incremental attribute reduction for dynamic data. Knowl.-Based Syst. **140**, 142–157 (2018)
75. Zhang, S., Guo, P., Zhang, J., Wang, X., Pedrycz, W.: A completeness analysis of frequent weighted concept lattices and their algebraic properties. Data Knowl. Eng. **81–82**, 104–117 (2012)
76. Zhang, W., Wei, L., Qi, J.: Attribute reduction theory and approach to concept lattice. Science in China Series F: Information Sciences **48**(6), 713–726 (2005)

Chapter 10
Towards Distributivity in FCA for Phylogenetic Data

Alain Gély, Miguel Couceiro, and Amedeo Napoli

10.1 Motivation

Lattices and median graphs are two structures with many applications, in particular in classification and knowledge discovery. Median graphs are especially used in biology, for example in phylogeny, for modeling inter-species relationships. In phylogeny, one of the main problems is to find evolution trees for representing existing species from accessible DNA fragments. When several trees are leading to the same inter-species phylogenetic relationship, the preferred ones are the most "parsimonious", where the number of modifications such as mutations for example, is minimal for the considered species. However, several possible parsimonious trees may exist simultaneously. Such a situation arises with inverse or parallel mutations, e.g., when a gene goes back to a previous state or the same mutation appears for two non-linked species. This calls for a generic representation of such a family of trees.

Bandelt et al. [2, 3] propose the notion of *median graph* to overcome this issue, since it was noticed that a median graph may encode all parsimonious trees. It is known that median graphs are related to lattices (see, e.g., [1, 3]). Any distributive lattice is a median graph, and any median graph can be thought of as a distributive $\wedge$-semilattice such that for all x, y, z if the supremum of each pair exists, then the supremum $\{x, y, z\}$ also exists.

Formal Concept Analysis (FCA) is based on lattice theory and can be used in classification and knowledge discovery. Uta Priss [15, 16] made a first attempt to use the algorithmic machinery of FCA and the links between distributive lattices and median graphs, to analyze phylogenetic trees. However, not every concept lattice is distributive, and thus FCA alone does not necessarily outputs median graphs. A

A. Gély (✉)
Université de Lorraine, CNRS, LORIA, Metz, France
e-mail: alain.gely@loria.fr

M. Couceiro • A. Napoli
Université de Lorraine, CNRS, Inria, LORIA, Nancy, France
e-mail: miguel.couceiro@loria.fr; amedeo.napoli@loria.fr

© The Author(s), under exclusive license to Springer Nature Switzerland AG 2022
R. Missaoui et al. (eds.), *Complex Data Analytics with Formal Concept Analysis*,
https://doi.org/10.1007/978-3-030-93278-7_10

transformation should be designed to build a median graph from a concept lattice. In [16] Uta Priss sketches an algorithm to convert any lattice into a median graph. The key step is to transform any lattice into a distributive lattice. However, how to transform a lattice into a distributive one is not detailed in this paper.

In [4], Bandelt uses a data set from [17] to illustrate and evaluate a median graph. In this introduction, we will re-use it to show the differences between median graph and FCA approaches. The example is an extract of mitochondrial DNA for 15 Kung individuals from a Khoisan-speaking hunter-gathered population in southern Africa. For some sequences in mitochondrial DNA (nucleotide positions, denoted by a, b, ... , j in table), a binary information indicates if a group of individuals owns the consensus version of the sequence (empty cell) or a variation for this sequence ($\times$ value). Eight individual groups are studied because some individuals share the same variants. For example, group 0 stands for 4 similar individuals in [4]. Individual group with no variation on any nucleotide positions (consensus group) is not shown on the table. These data are shown in Fig. 10.1a.

For these data, the median graph is shown in Fig. 10.1b. Vertices are either individual group (numbered from 0 to 7) or latent vertices (spotted as Lx), added such that from a group to an adjacent one, only one variation exists for nucleotide sequences (principle of parsimony supposes that there is no chance that two variations arise exactly at the same moment for the same population in evolutionary process). This variation is indicated on edges. As an example, from consensus group to vertex $L4$, the only variation occurs in sequence k, from vertex $L4$ to vertex 0 the only variation occurs in sequence j.

As stated, this graph contains every parsimonious tree as covering tree. Median graph owns others good properties: remove edges labeled with a sequence variation produces two disconnected parts. One corresponds to individuals with the variation, the other without the variation. For example, with edges labelled j, one can partition the vertices in two connected components: vertices $L1, 0, 2, 4$ and 7 are individuals with a variant on position j. Vertices in the other connected component represent individuals without this variant.

For individuals represented by vertex 0, there is not enough data to determine if the evolutionary path from the consensus started with a variant on k position (giving latent vertex $L4$) or started with a variant on j (giving latent vertex $L1$)? Strength of median graph is to display these two possibilities, which is not possible with a parsimonious tree.

Since data is a binary table, Formal Concept Analysis can be applied. The concept lattice obtained from the data is shown in Fig. 10.1c. In general, it does not correspond to a median graph. To build a median graph, a necessary condition is to have a distributive $\vee$-semilattice. In [11], based on the work of Birkhoff and FCA formalism, we propose an algorithm to compute such a semilattice (and the corresponding data table). The result of this algorithm, transforming a concept lattice into a median graph, is given in Fig. 10.1d. Since FCA is supported by a wide community, the main idea of this research is to be able to use FCA results and software to deal with phylogenetic data and median graphs.

Remark that, for this particular data set the algorithm computes the median graph obtained by Bandelt in [4]. Unfortunately, in some cases the algorithm returns a distributive ∨-semilattice L_d that is not minimal: There exists $L_{d'}$ such that L can be embedded in $L_{d'}$ and $L_{d'}$ can be embedded L_d.

The continuation of [11] is to search for an algorithm which outputs a minimal distributive ∨-semilattice. Since we look for minimality, a natural question arises: does a unique minimal distributive ∨-semilattice L_d exist? In this paper, we propose a counter-example which shows that a minimum distributive ∨-semilattice does not always exist.

In the following section, we recall definitions and notation for the understanding of this paper. We then sketch the limitations of our algorithm and show in Sect. 10.4 that a minimum distributive ∨-semilattice does not exist. We conclude this paper by some remarks and perspectives in Sect. 10.5.

10.2 Models: Lattices, Semilattices, Median Algebras and Median Graphs

In this section we recall basic notions and notation needed throughout the paper. We will mainly adopt the formalism of [10], and we refer the reader to [8, 9] for further background. *In this paper, all sets are supposed to be finite.*

10.2.1 Lattices and FCA

A *partially ordered set* (or *poset* for short) is a pair $(P, \leq)$ where P is a set and $\leq$ is a *partial order* on P, that is, a reflexive, antisymmetric and transitive binary relation on P.

An upper (*resp. lower*) bound of $X \subseteq P$ is an element $y \in P$ such that $\forall x \in X$, $x \leq y$ (*resp* $y \leq x$). For $X \in P$, the lowest upper bound, if it exists, is called the join or the supremum. The greatest lower bound, if it exists, is called the meet or the infimum.

A ∨-semilattice $(L, \leq)$ is an ordered set such that the supremum exists for all $X \subseteq L$, $X \neq \emptyset$. For $x, y \in L$, $x \vee y$ denotes the supremum. Such an order has a lowest element (bottom) denoted by $\bot = \bigwedge L$. A ∧-semilattice $(L, \leq)$ is an ordered set such that the infimum exists for all $X \subseteq L$. For $x, y \in L$, $x \wedge y$ denotes the infimum. Such an order owns a greatest element (top) denoted by $\top = \bigvee L$. A lattice $(L, \leq)$ is an ordered set such that a supremum and an infimum exist for all $X \subseteq L$.

An element $x \in L$ such that $x = y \vee z$ implies $x = y$ or $x = z$ is called a ∨-irreducible element. Dually, an element $x \in L$ such that $x = y \wedge z$ implies $x = y$ or $x = z$ is called a ∧-irreducible element. We will denote the set of ∧-irreducible elements and ∨-irreducible elements of L by $\mathscr{M}(L)$ and $\mathscr{J}(L)$, respectively. Observe that both $\mathscr{M}(L)$ and $\mathscr{J}(L)$ are posets when ordered by $\leq$.

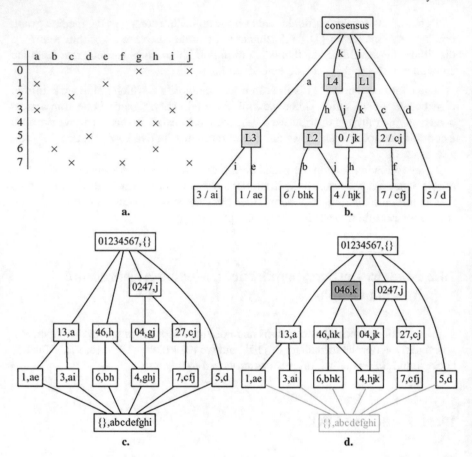

Fig. 10.1: (**a**) Phylogenetic data of sequence variations for individual groups. (**b**) Median graph obtained from the data [4]. (**c**) Concept lattice for phylogenetic data. (**d**) Concept lattice corresponding to the median graph. The new concept $(046, k)$ corresponds to $L4$. To obtain this lattice, data must be modified replacing column g by k

Posets and lattices can be represented and visualized by their Hasse-diagrams [9]. Examples of lattices are given in Figs. 10.2 and 10.3. N_5 and M_3 lattices are involved in non distributivity (see Sect. 10.2.2). In Figs. 10.2 and 10.3, $\vee$-irreducible elements are labeled with numbers and $\wedge$-irreducible elements are labeled with letters. Some elements, doubly irreducible, have two labels. This is the case in Fig. 10.3.c for element which is labeled either by 1 and a and for element which is labeled either by 3 and c. Elements with labels d and e are $\wedge$-irreducible (but no $\vee$-irreducible). Element with label 2 is $\vee$-irreducible (but no $\wedge$-irreducible). Formal Concept Analysis [10] uses *concept lattices* for data analysis tasks. A concept lattice

$(\mathscr{J}(L),\mathscr{M}(L),\leq)$	a	b	c
1	×	×	
2		×	
3			×

a.

$(\mathscr{J}(L),\mathscr{J}(L),\not\geq)$	$1(c)$	$2(d)$	$3(b)$
1		×	×
2			×
3	×	×	

b.

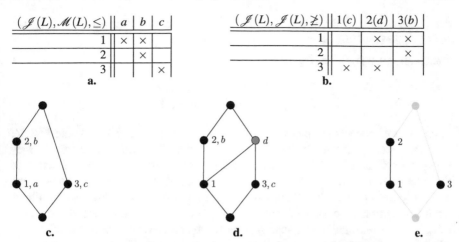

c.   **d.**   **e.**

Fig. 10.2: **(a)** Standard context for lattice N_5 **(b)** The context $(\mathscr{J}(N_5),\mathscr{J}(N_5),\not\geq)$ of an ideal (and so, distributive) lattice. **(c)** N_5, a non distributive lattice. **(d)** Ideal lattice for $(\mathscr{J}(N_5),\leq)$. **(e)** $(\mathscr{J}(L),\leq)$ poset for ideal lattice in **(d)**. Note that N_5 can be order-embedded in the lattice . Concepts (X,Y) are maximal rectangles of the contexts. For an element e of the lattice, the corresponding concept (X,Y) is $X = \{j \in \mathscr{J}(L) \mid j \leq e\}$ and $Y = \{m \in \mathscr{M}(L) \mid m \geq e\}$. For example, the element with label d is the concept $(\{1,3\},\{d\})$

$(\mathscr{J}(M_3),\mathscr{M}(M_3),\leq)$	a	b	c
1	×		
2		×	
3			×

a.

$(\mathscr{J}(L),\mathscr{M}(L),\leq)$	a	c	d	e
1	×		×	
2			×	×
3		×		×

b.

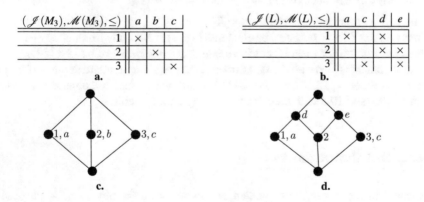

c.   **d.**

Fig. 10.3: **a** the standard context for lattice M_3. **b** the standard context $(\mathscr{J}(L),\mathscr{M}(L),\leq)$ for L in Fig. 10.3.d **c** M_3 is a non distributive lattice. **d** A non distributive lattice L such that $(\mathscr{J}(L),\leq) = (\mathscr{J}(M_3),\leq)$

is built from a binary table, which is called a *formal context*, thanks to the prime (′) operators.

We denote by (G,M,I) a formal context where G is a set of objects, M a set of attributes and I an incidence relation between objects and attributes. In phylogenetic data, objects are usually species, attributes are mutations, and $(g,m) \in I$—or gIm—indicates that mutation m is spotted in species g.

Definition 10.1 For a set $X \subseteq G$, $Y \subseteq M$ we define:

$$X' = \{y \in M \mid xIy \text{ for all } x \in X\}$$
$$Y' = \{x \in G \mid xIy \text{ for all } y \in Y\}$$

Then a formal concept is a pair (X,Y) where $X \subseteq G$, $Y \subseteq M$ and $X' = Y$ and $Y' = X$. X is the extent and Y is the intent of the concept. They are closed sets as they verify $X = X''$ and $Y = Y''$. The set of all formal concepts ordered by inclusion of the extents—dually the intents—denoted by $\leq$ generates the *concept lattice* of the context (G,M,I). The existence of a supremum and an infimum allows to use lattices for classification process. Concepts can be viewed as classes, indeed a concept (X,Y) is a representation of a maximal set of objects X which share a maximal set of attributes Y. If another concept (X_1,Y_1) is greater than (X,Y), it contains more objects, but described by fewer attributes. X_1 is a class, more general than X.

A *clarified context* is a context such that $x' = y'$ implies $x = y$ for any element of G and any element of M. In a clarified context, the set of attributes of two distinct objects are distinct, and dually for objects. Moreover, a clarified context is *reduced* iff it does not contain:

- a vertex $g \in G$ such that $g' = X'$ with $X \subseteq G$, $g \notin X$
- a vertex $m \in M$ such that $m' = Y'$ with $Y \subseteq M$, $m \notin Y$

Indeed, a vertex $x \in G$ such that $x' = X'$ with $X \subseteq G$, $x \notin X$ is a reducible element (since it intent may be obtained by elements in the set X). Only irreducible elements, which are not join or meet of other elements, are necessary to build a lattice. The reduced context is also called a *standard context* [10]. The standard context of lattice L is such that $G = \mathscr{J}(L)$ and $M = \mathscr{M}(L)$. Examples of standard contexts are given in Figs. 10.2 and 10.3 with their corresponding concept lattice.

10.2.2 Distributive Lattices

As stated in the motivations, median graphs are used for phylogenetic purposes, and encode a family of trees. It is known that theses graphs can be considered as particular distributive $\vee$-semilattices. This subsection provides basic notions about distributive (semi)lattices.

A lattice is *distributive* if $\wedge$ and $\vee$ are distributive with respect to each other. Formally, a lattice L is distributive if for every $x,y,z \in L$, we have that one (or, equivalently, both) of the following identities holds:

$$(i)\; x \vee (y \wedge z) = (x \vee y) \wedge (x \vee z), \quad (ii)\; x \wedge (y \vee z) = (x \wedge y) \vee (x \wedge z).$$

Distributive lattices appear a natural choice for any classification task or as computation and semantic models; see, e.g., [8, 9, 12, 13]. This is partially due to the fact that *any distributive lattice can be thought of as a sublattice of a power-set lattice, i.e., the set $\mathscr{P}(X)$ of subsets of a given set X.*

Note that the definition of a sublattice is more constraint than the definition of a suborder: A subset $X \subseteq L$ is a *sublattice* of L if for every $x, y \in X$ we have that $x \wedge y, x \vee y \in X$. For example, N_5 (Fig. 10.2c) is a suborder of the lattice in Fig. 10.2d, but is not a sublattice. Indeed, $1 \vee 3$ is not the same element in the two lattices.

The distributivity property of lattices has been equivalently described in several ways. One of these properties relies on the notion of sublattice, as follows:

Property 10.1 L is a distributive lattice iff it contains neither N_5 nor M_3 as sublattices.

This property describes distributive lattices in terms of two forbidden structures, namely, M_3 and N_5 that are, up to isomorphism, the smallest non distributive lattices. N_5 is represented in Fig. 10.2c and M_3 is represented in Fig. 10.3c. In Fig. 10.2d the lattice does not contain N_5 nor M_3 as sublattices, and so is distributive. In Fig. 10.3d the lattice does not contain M_3 as sublattice, but contains N_5 as a sublattice, and so, is not distributive.

Our goal is to transform a lattice into a distributive one. For this particular task, the Birkhoff's representation of distributive lattice is of practical interest. It uses the notion of *order ideal*, recalled here:

Definition 10.2 (Order Ideal) Let $(P, \leq)$ be a poset. For a subset $X \subseteq P$, let $\downarrow X = \{y \in P : y \leq x \text{ for some } x \in X\}$ and $\uparrow X = \{y \in P : x \leq y \text{ for some } x \in X\}$. A set $X \subseteq P$ is a *(poset) ideal* (resp. *filter*) if $X = \downarrow X$ (resp. $X = \uparrow X$). If $X = \downarrow \{x\}$ (resp. $X = \uparrow \{x\}$) for some $x \in P$, then X is said to be a *principal* ideal (resp. filter) of P. For principal ideals, we omit brackets, so that $\uparrow x$ (resp. $\downarrow x$) stands for $\uparrow \{x\}$ (resp. $\downarrow \{x\}$)

$(\mathscr{J}(L), \mathscr{J}(L), \not\geq)$	$1(f)$	$2(e)$	$3(d)$
1		×	×
2	×		×
3	×	×	

a.

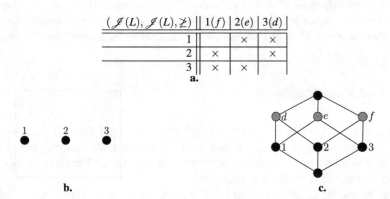

b.   **c.**

Fig. 10.4: **a** The context $(\mathscr{J}(L), \mathscr{J}(L), \not\geq)$ for $(J(L), \leq)$ the poset induced by $\vee$-irreducible elements of lattices in Fig. 10.3. **b** $(\mathscr{J}(L), \leq)$ for L (or equivalently M_3 in Fig. 10.3). **c** Ideal lattice for $(\mathscr{J}(L), \leq)$ (equivalently $(\mathscr{J}(M_3), \leq)$). M_3 and L can be order-embedded in this lattice

Birkhoff's Representation of Distributive Lattices. Let $(P, \leq)$ be a poset and consider the set $\mathscr{O}(P)$ of ideals of P, i.e.,

$$\mathscr{O}(P) = \{\bigcup_{x \in X} \downarrow x \mid X \subseteq P\}.$$

It is well-known that for every poset P, the set $\mathscr{O}(P)$ ordered by inclusion is a distributive lattice, called *ideal lattice* of P [5, 6]. Furthermore, the poset of $\vee$-irreducible elements of $\mathscr{O}(P)$ is $\mathscr{J}((\mathscr{O}(P))) = \{\downarrow x \mid x \in P\}$ and it is (order) isomorphic to P.

This representation is used to provide a distributive lattice L_d with the same poset of $\vee$-irreducible elements as an arbitrary lattice L. In this case, L is order-embedded in L_d. For example, in Fig. 10.2d, the lattice is the ideal lattice of $(\mathscr{J}(N_5), \leq)$. In the same way, the two lattices in Figs. 10.3c and 10.3d have the same poset of $\vee$-irreducible elements, and can be order embedded in the ideal lattice of this poset, represented in Fig. 10.4. In particular, in [7, 14] it is shown that the family of lattices with the same poset of $\vee$-irreducible elements is itself a lattice, an so there exists a minimum element.

From a poset P, it is possible to obtain the context of the ideal lattice as $C_{ideal(P)} = (P, P, \not\geq)$. For a standard context $C = (\mathscr{J}(L), \mathscr{M}(L), \leq)$, the standard context of the ideal lattice of $(\mathscr{J}(L), \leq)$ is $C_{ideal(J)} = (\mathscr{J}(L), \mathscr{J}(L), \not\geq)$. Note also that, for every distributive lattice L, the two posets $\mathscr{J}(L)$ and $\mathscr{M}(L)$ are dually-isomorphic. This is why the standard contexts of distributive lattices are "squares" ($|\mathscr{J}(L)| = |\mathscr{M}(L)|$), and are built with the information of only one of these two posets.

In the following subsection, we give some hints about median graphs and median algebras. As we will soon observe, the class of median graphs is in correspondence with a particular subclass of distributive $\vee$-semilattices.

Let L be a $\vee$-semilattice and $x \in L$, then $\uparrow x$ is a lattice (in the finite case, every $\vee$-semilattice with a lowest element is a lattice). Then a $\vee$-semilattice L is distributive iff $\uparrow x$ is distributive, for all x [8]. In practice, it is sufficient to check this property only for minimal elements of L. Indeed, filters of non minimal elements are sublattices of a minimal element filter, and sublattices of distributive lattices are distributive.

10.2.3 Median Graphs

As said in the introduction, a median graph encodes all parsimonious phylogenetic trees. A median graph is a connected graph having the median property, i.e. for any three vertices a, b, c, there is exactly one vertex x which lies on a shortest path between each pair of vertices in $\{a, b, c\}$.

The following characterization of distributive lattices explains some links of distributive lattice with median graphs and median algebras.

Property 10.2 A lattice L is a distributive lattice iff for all $x, y, z \in L$,

$$(x \wedge y) \vee (y \wedge z) \vee (z \wedge x) = (x \vee y) \wedge (y \vee z) \wedge (z \vee x).$$

This property establishes a correspondence between distributive lattices and median algebras. Indeed, a median algebra is a structure (M,m) where M is a nonempty set and $m : M^3 \to M$ is an operation, called *median operation*, that satisfies the following conditions $m(a,a,b) = a$ and $m(m(a,b,c),d,e) = m(a,m(b,c,d),m(b,c,e))$, for every $a,b,c,d,e \in M$. It is not difficult to see that if L is distributive, then $m(a,b,c) = (a \wedge b) \vee (b \wedge c) \vee (c \wedge a)$ is a median operation. The connection to *median graphs* was established by Avann [1] who showed that every median graph is the Hasse diagram of a median algebra (thought of as a semilattice). For further background on median structures see, e.g., [3]. This result was later used by Bandelt [2] to establish the link between distributive lattices and median graphs.

Property 10.3 A graph is a median graph iff it is isomorphic to a $\vee$-semilattice L with the two following properties:

- L is distributive
- for all $x,y,z \in L$ such that $(x \wedge y)$, $(y \wedge z)$ and $(z \wedge x)$ are defined, $(x \wedge y \wedge z)$ is defined.

10.3 Algorithm to Produce a Distributive $\vee$-Semilattice

To build a median graph from a context using FCA, a necessary condition is to build a distributive $\vee$-semilattice. For the concept lattice L, it is always possible to consider the semilattice $L_\vee = L \backslash \bot$ (L minus the lowest element). Minimal elements of this semilattice $L_\vee$ are minimal elements of $(\mathscr{J}(L), \leq)$.

It remains to transform the filter of these elements into a distributive lattice. Our previous work [11] is based on Birkhoff's representation of a distributive lattice. Since sublattices of a distributive lattice are distributive, a simple way to obtain a distributive $\vee$-semilattice from a lattice L is to map L into the ideal lattice of $(J(L), \leq)$. In practice, the bottom element exists because of the existence of infimum in lattice, but it usually does not carry semantic information for classification. For example, trees are median graphs and so distributive $\vee$-semilattice (considering the root as the greatest element) but obviously not lattices. With the addition of a bottom element $\bot$, the trees become lattices. There is no reason that these lattices are distributive. Two trivial examples are N_5 and M_3: Once the lowest element is removed, either $N_{5\vee} = N_5 \backslash \bot$ and $M_{3\vee} = M_3 \backslash \bot$, considered as $\vee$-semilattices, are distributive (N_5 and M_3 are isomorphic to path and tree). Nevertheless, neither N_5 nor M_3 are distributive.

Now, the mapping of the concept lattice into its ideal lattice will produce a $\vee$-semilattice, but this is not necessarily a minimal solution. For example, M_3 will be embedded in the Boolean lattice while $M_3 \backslash \bot$ is already a distributive $\vee$-semilattice. Hence, the global approach that embeds a concept lattice into its ideal lattice is not efficient.

Alternatively, we can think of a local approach: instead of embedding the whole concept lattice into its ideal lattice, we do so for the sublattices corresponding to filters of minimal elements of $\mathscr{J}(L)$. The algorithm proposed in [11] computes contexts of $\uparrow j$ for every minimal $\vee$-irreducible element j, and transforms these contexts so that they correspond to the context of a distributive lattice. Once these contexts are built, we merge them to build the whole lattice. However, this method

---

Algorithm 5: Construction of context of a distributive $\vee$-semilattice

---

Data: A context $(\mathscr{J}(L), \mathscr{M}(L), I)$ of a lattice L
Result: the context $(\mathscr{J}(L_{med}), \mathscr{M}(L_{med}), I)$ of a distributive $\vee$-semilattice L_{med} such that L can be order-embedded in L_{med}

1 **foreach** $j \in \mathscr{J}(L)$, *minimal* **do**
2    $(P_j, \le) \leftarrow \emptyset$

3 **repeat**
4   stability $\leftarrow$ true; **foreach** $j \in \mathscr{J}(L)$, *minimal* **do**
5   compute P_j the poset of $\vee$-irreducible elements in $\uparrow j$
6   compute $C_j = (P_j, P_j, \not\ge)$
7   **if** P_j *modified since last iteration* **then**
8   stability $\leftarrow$ false;

9   Merge all $C_j = (P_j, P_j, \not\ge)$ in a unique context
10   Reduce this context
11 **until** *stability*

---

does not always output a minimal solution, i.e., there may exist $L_{d'}$ a distributive $\vee$-semilattice such that L can be embedded in $L_{d'}$, and $L_{d'}$ can be embedded in L_d with $|L| < |L_{d'}| < |L_d|$ and $(\mathscr{J}(L), \le) = (\mathscr{J}(L_{d'}), \le) = (\mathscr{J}(L_d), \le)$. This result comes from the fact that each filter is processed independently. Nevertheless, it is possible that some elements are shared by several filters of minimal $\vee$-irreducible elements. This is illustrated in Fig. 10.5, and motivates the two following observations. First, it is possible that some elements added to a filter for achieving distributivity belong to other filters. These new elements may break a previously obtained distributivity in others filters. This is the case in Fig. 10.5b with the two new elements r and g. At a first iteration of the loop of the algorithm, when the filters are merged, r and g are distinct elements. Neither the filter of 1 nor 2 are distributive. $\uparrow 1$ (resp. $\uparrow 2$) is not distributive because of r (resp. g). To overcome the problem, the algorithm loops while any filter is modified by the process. At worst, the algorithm computes the context corresponding to the ideal lattice of $\vee$-irreducible poset of L and the algorithm always terminates.

Second, in some cases, a minimal solution cannot be reached when locally considering the filters. Such a solution is proposed if Fig. 10.5d.

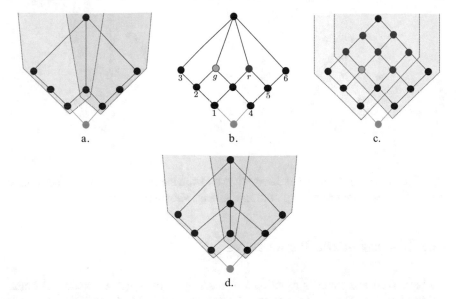

Fig. 10.5: **a** A lattice, with its two atom filters enlightened. **b** Result of the first step of the algorithm. **c** Result of the algorithm. **d** A minimal distributive ∨-semilattice (not reachable by the algorithm)

10.4 A Counter-Example for the Existence of a Minimum Distributive ∨-Semilattice

The local approach thus seems to be better than the global one. However, our algorithm does not always produce a minimal solution. The natural question is then whether, for a lattice L, there exists a minimum (i.e., minimal and unique) distributive ∨-semilattice L_d such that L can be embedded into L_d. We will now show through a counter-example that such minimum does not always exist.

The proposed counter-example is given in Fig. 10.6: For the lattice shown in Fig. 10.6a, either lattice in Figs. 10.6b and 10.6c are minimal distributive ∨-semilattices (since they differ by one element only) but it is obvious that lattices in Fig. 10.6b and 10.6c are not isomorphic. So, since a minimum solution does not exist, some choices remain to be done in order to use FCA algorithms for traditional application fields of median graphs, in particular for phylogeny.

Since phylogenetic parsimonious trees minimize the number of required evolutions to obtain species, lattice in Fig. 10.6b may be preferred to lattice in Fig. 10.6c In effect, in the first case, vertex b is obtained from a mutation at the root, in the latter, two mutations from the root are needed.

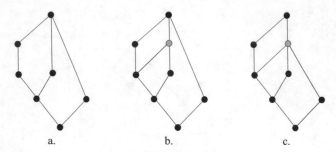

Fig. 10.6: (**a**) A lattice such that there exists two non isomorphic minimal distributive ∨-semilattices (when removing bottom element) shown in (**b**) and (**c**)

10.5 Discussion and Perspectives

We have seen that there is a lattice L for which there is not a unique minimum distributive ∨-semilattice L_d such that L can be embedded in L_d and with the same posets of ∨-irreducible elements $((\mathcal{J}(L), \leq) = (\mathcal{J}(L_d), \leq))$.

So, even if we provide an algorithm that produces a minimal solution, the question of the meaning of this (not unique) solution should be addressed. A way to tackle it is to find an algorithm able to list all the minimal solutions. Alternatively, we could propose a measure of "interestingness" of these minimal solutions, so that an optimal solution could be reached based on such a measure. This remains a topic of current research.

Also, this work was motivated by the study of the relations between distributive ∨-semilattices and median graphs. Not all distributive ∨-semilattices are median graphs. It remains to check the following condition: For every triple of elements x, y, z such that $x \wedge y$, $x \wedge z$ and $y \wedge z$ are defined, $x \wedge y \wedge z$ is defined.

It is obvious that this condition is not satisfied for some distributive ∨-semilattices. A trivial example is the Boolean lattice (minus the bottom element) but in this particular case, the whole lattice is distributive, and so a median graph. Nevertheless, it remains open whether this is always the case.

References

1. Avann, S.P.: Median algebras. Proceedings of the American Mathematical Society **12**, 407–414 (1961)
2. Bandelt, H.J., Forster, P., Röhl, A.: Median-joining networks for inferring intraspecific phylogenies. Molecular biology and evolution **16**(1), 37–48 (1999)
3. Bandelt, H.J., Hedlíková, J.: Median algebras. Discrete mathematics **45**(1), 1–30 (1983)
4. Bandelt, H.J., Macaulay, V., Richards, M.: Median networks: speedy construction and greedy reduction, one simulation, and two case studies from human mtdna. Molecular phylogenetics and evolution **16**(1), 8–28 (2000)
5. Birkhoff, G.: Rings of sets. Duke Math. J. **3**(3), 443–454 (1937).

6. Birkhoff, G., Frink, O.: Representations of lattices by sets. Transactions of the American Mathematical Society **64**(2), 299–316 (1948)
7. Bordalo, G.H., Monjardet, B.: The lattice of strict completions of a poset. Electronic Notes in Discrete Mathematics **5**, 38–41 (2000)
8. Caspard, N., Leclerc, B., Monjardet, B.: Finite ordered sets: concepts, results and uses. Cambridge University Press (2012)
9. Davey, B.A., Priestley, H.A.: Introduction to Lattices and Order. Cambridge university press (2002)
10. Ganter, B., Wille, R.: Formal Concept Analysis: Mathematical Foundations. Springer (1999)
11. Gély, A., Couceiro, M., Napoli, A.: Steps towards achieving distributivity in formal concept analysis. In: Proceedings of the Fourteenth International Conference on Concept Lattices and Their Applications, CLA 2018, Olomouc, Czech Republic, June 12–14, 2018., pp. 105–116 (2018).
12. Hopcroft, J.E., Motwani, R., Rotwani, Ullman, J.D.: Introduction to Automata Theory, Languages and Computability, 2nd edn. Addison-Wesley Longman Publishing Co., Inc., Boston, MA, USA (2000)
13. Mattern, F.: Virtual time and global states of distributed systems. Parallel and Distributed Algorithms **1**(23), 215–226 (1989)
14. Nation, J., Pogel, A.: The lattice of completions of an ordered set. Order **14**(1), 1–7 (1997)
15. Priss, U.: Concept lattices and median networks. In: CLA, pp. 351–354 (2012)
16. Priss, U.: Representing median networks with concept lattices. In: ICCS, pp. 311–321. Springer (2013)
17. Vigilant, L., Pennington, R., Harpending, H., Kocher, T.D., Wilson, A.C.: Mitochondrial dna sequences in single hairs from a southern african population. Proceedings of the National Academy of Sciences **86**(23), 9350–9354 (1989)

Chapter 11
Triclustering in Big Data Setting

Dmitry Egurnov, Dmitry I. Ignatov, and Dmitry Tochilkin

11.1 Introduction

Mining of multimodal patterns in n-ary relations or Boolean tensors is among popular topics in Data Mining and Machine Learning [3, 9, 14, 15, 28, 37, 38]. Thus, cluster analysis of multimodal data and specifically of dyadic and triadic relations is a natural extension of the idea of original clustering. In dyadic case, biclustering methods (the term bicluster was coined in [29]) are used to simultaneously find subsets of objects and attributes that form homogeneous patterns of the input object-attribute data. Triclustering methods operate in triadic case, where for each object-attribute pair one assigns a set of some conditions [10, 17, 30]. Both biclustering and triclustering algorithms are widely used in such areas as gene expression analysis [6, 22, 26, 27, 44], recommender systems [18, 20, 32], social networks analysis [12], natural language processing [39], etc. The processing of numeric multimodal data is also possible by modifications of existing approaches for mining binary relations [21]. Another interesting venue closely related to attribute dependencies in object-attribute data also take place in triadic case, namely, mining of triadic association rules and implications [7, 31].

Though there are methods that can enumerate all triclusters satisfying certain constraints [3] (in most cases they ensure that triclusters are dense), their time complexity is rather high, as in the worst case the maximal number of triclusters is usually exponential (e.g. in case of formal triconcepts), showing that these methods are hardly scalable. Algorithms that process big data should have at most linear time complexity (e.g., $O(|I|)$ in case of n-ary relation I) and be easily parallelisable. In addition, especially in the case of data streams [36], the output patterns should be the results of one pass over data.

Earlier, in order to create an algorithm satisfying these requirements, we adapted a triclustering method based on prime operators (prime OAC-triclustering method)

D. Egurnov (✉) • D. I. Ignatov • D. Tochilkin
National Research University Higher School of Economics, Moscow, Russia
e-mail: egurnovd@yandex.ru; dignatov@hse.ru; dstochilkin@gmail.com

[10] and proposed its online version, which has linear time complexity; it is also one-pass and easily parallelisable [11]. However, its parallelisation is possible in different ways. For example, one can use a popular framework for commodity hardware, Map-Reduce (M/R) [35]. In the past, there were several successful M/R implementations in the FCA community and other lattice-oriented domains. Thus, in [23], the authors adapted Close-by-One algorithm to M/R framework and showed its efficiency. In the same year, in [24], an efficient M/R algorithm for computation of closed cube lattices was proposed. The authors of [42] demonstrated that iterative algorithms like Ganter's NextClosure can benefit from the usage of iterative M/R schemes.

Our previous M/R implementation of the triclustering method based on prime operators was proposed in [45] showing computational benefits on rather large datasets. M/R triclustering algorithm [45] is a successful distributed adaptation of the online version of prime OAC-triclustering [11]. This method uses the MapReduce approach as means for task allocation on computational clusters, launching the online version of prime OAC-triclustering on each reducer of the first phase. However, due to the simplicity of this adaptation, the algorithm does not use the advantages of MapReduce to the full extent. In the first stage, all the input triples are split into the number of groups equals to the number of reducers by means of hash-function for entities of one of the types, object, attribute, or condition, which values are used as keys. It is clear that this way of data allocation cannot guarantee uniformness in terms of group sizes. The JobTracker used in Apache Hadoop is able to evenly allocate tasks by nodes.[1] To do so, the number of tasks should be larger than the number of working nodes, which is not fulfilled in this implementation. For example, let us assume we have 10 reduce SlaveNodes; respectively, $r = 10$, and the hash function is applied to the objects (the first element in each input triple). However, due to the non-uniformity of hash-function values by modulo 10, it may happen that the set of objects will result in less than 10 different residuals during division by 10. In this case, the input triples will be distributed between parts of different sizes and processed by only a part of cluster nodes. Such cases are rather rare; it could be possible only for slicing by entities (objects, attributes, or conditions) with a small number of different elements. However, they may slow down the cluster work drastically.

The weakest link is the second stage of the algorithm. First of all, during the first stage, it finds triclusters computed for each data slice separately. Hence, they are not the final triclusters; we need to merge the obtained results.

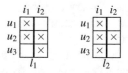

Table 11.1: An example with triadic data

---
[1] https://wiki.apache.org/hadoop/JobTracker.

Let us consider an example in Table 11.1 with the ternary relation on users-items-labels. Let us assume that the first mapper splits data according to their labels' component, $r = 2$, then triples containing label l_1 and those related to label l_2 are processed on different nodes. After the first stage completion on the first node, we have tricluster $(\{u_2\}, \{i_1, i_2\}, \{l_1\})$ among the others, while the second node results in tricluster $(\{u_2\}, \{i_1, i_2\}, \{l_2\})$. It is clear that both triclusters are not complete for the whole input dataset and should be merged into $(\{u_2\}, \{i_1, i_2\}, \{l_1, l_2\})$. The second stage of the algorithm is responsible for this type of merging. However, as one can see, this merging assumes that all intermediate data should be located on the same node. In a big data setting, this allocation of all the intermediate results on a single node is a critical point for application performance.

To calculate tricluster components (or cumuli, see Sect. 11.3.1) and assemble the final triclusters from them, we need to have large data slices on the computational nodes. Moreover, during parallelisation of the algorithm, those data slices can be required simultaneously on different nodes. Thus, to solve these problems one needs to fulfil data centralisation (all the required slices should be present at the same node simultaneously). However, it leads to the accumulation of too large parts of the data as described. Another approach is to perform data replication. In this case, the total amount of data processed on computational nodes is increased, while the data are evenly distributed in the system. The latter approach has been chosen for our updated study on multimodal clustering.

Note that experts warn the potential M/R users: "the entire distributed-file-system milieu makes sense only when files are very large and are rarely updated in place" [35]. In this work, as in our previous study, we assume that there is a large bulk of data to process that is not coming online. As for experimental comparison, since we would like to follow an authentic M/R approach, without using the online algorithm [11], we implement and enhance only MapReduce schemes from the appendix in [45] supposed for future studies in that time.

The rest of the paper is organised as follows: in Sect. 11.2, we recall the original method and the online version of the algorithm of prime OAC-triclustering. Section 11.3 is dedicated to extensions of OAC triclustering. Specifically, Sect. 11.3.1 generalise prime OAC-triclustering for the case of multimodal data, and Sect. 11.3.2 touches the case of multi-valued context. In Sect. 11.4 we describe implementation details. Section 11.4.1, gives the M/R setting of the problem and the corresponding M/R version of the original algorithm with important implementation aspects. In Sect. 11.4.3 we speak about parallel implementation of OAC triclustering. Finally, in Sect. 11.5 we show the results of several experiments that demonstrate the efficiency of the M/R version of the algorithm.

11.2 Prime Object-Attribute-Condition Triclustering

Prime object-attribute-condition triclustering method (OAC-prime) based on triadic version of Formal Concept Analysis [8, 25, 41] is an extension for the triadic case of object-attribute biclustering method [16]. The rationale for introduction of this type of triadic patterns is as follows: OAC-triclusters provide a reasonable approximation for formal triconcepts [15, 17] and thus continues development of OA-biclustering as another fault-tolerant approximation of formal concepts in dyadic case [2, 16].

Triclusters generated by this method have a similar structure as object-attribute biclusters, namely the cross-like structure of triples inside the input data cuboid (i.e. formal tricontext).

Let $\mathbb{K} = (G, M, B, I)$ be a triadic context , where G, M, B are respectively the sets of objects, attributes, and conditions, and $I \subseteq G \times M \times B$ is a triadic incidence relation. Each prime OAC-tricluster is generated by applying the following prime operators to each pair of components of some triple:

$$(X,Y)' = \{b \in B \mid (g,m,b) \in I \text{ for all } g \in X, m \in Y\},$$
$$(X,Z)' = \{m \in M \mid (g,m,b) \in I \text{ for all } g \in X, b \in Z\}, \tag{11.1}$$
$$(Y,Z)' = \{g \in G \mid (g,m,b) \in I \text{ for all } m \in Y, b \in Z\},$$

where $X \subseteq G, Y \subseteq M$, and $Z \subseteq B$.

Then the triple $T = ((m,b)', (g,b)', (g,m)')$ is called *prime OAC-tricluster* based on triple $(g,m,b) \in I$. The components of tricluster are called, respectively, *tricluster extent, tricluster intent*, and *tricluster modus*. The triple (g,m,b) is called a *generating triple* of the tricluster T. The density of a tricluster $T = (G_T, M_T, B_T)$ is the ratio of actual number of triples in the tricluster to its size:

$$\rho(T) = \frac{|G_T \times M_T \times B_T \cap I|}{|G_T||M_T||B_T|} .$$

In these terms a triconcept is a tricluster with density $\rho = 1$.

Figure 11.1 shows the structure of an OAC-tricluster (X,Y,Z) based on triple $(\widetilde{g}, \widetilde{m}, \widetilde{b})$, triples corresponding to the gray cells are contained in the tricluster, other triples may be contained in the tricluster (cuboid) as well.

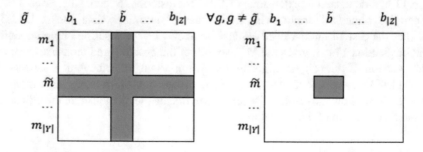

Fig. 11.1: Structure of prime OAC-triclusters: the dense cross-like central layer containing $\tilde{g}$ (left) and the layer for an object g (right) in $M \times B$ dimensions

The basic algorithm for the prime OAC-triclustering method is rather straightforward (see [10]). First of all, for each combination of elements from each of the two sets of $\mathbb{K}$ we apply the corresponding prime operator (we call the resulting sets *prime sets*). After that, we enumerate all triples from I and on each step, we must generate a tricluster based on the corresponding triple, check whether this tricluster is already contained in the tricluster set (by using hashing) and also check extra conditions.

The total time complexity of the algorithm depends on whether there is a non-zero minimal density threshold or not and on the complexity of the hashing algorithm used. In case we use some basic hashing algorithm processing the tricluster's extent, intent and modus without a minimal density threshold, the total time complexity is $O(|G||M||B| + |I|(|G| + |M| + |B|))$ (assuming the hashing algorithm takes $O(1)$ to operate, it takes $O(|G||M||B|)$ to precompute the prime sets and then $O(|G| + |M| + |B|)$ for each triple $(g, m, b) \in I$ to compute the hash); in case of a non-zero minimal density threshold, it is $O(|I||G||M||B|)$ (since computing density takes $O(|G||M||B|)$ for each tricluster generated from a triple $(g, m, b) \in I$). The memory complexity is $O(|I|(|G| + |M| + |B|))$, as we need to keep the dictionaries with the prime sets in memory.

In online setting, for triples coming from triadic context $\mathbb{K} = (G, M, B, I)$, the user has no a priori knowledge of the elements and even cardinalities of G, M, B, and I. At each iteration, we receive some set of triples from I: $J \subseteq I$. After that, we must process J and get the current version of the set of all triclusters. It is important in this setting to consider every pair of triclusters as being different as they have different generating triples, even if their respective extents, intents, and modi are equal. Thus, any other triple can change only one of these two triclusters, making them different.

To efficiently access prime sets for their processing, the dictionaries containing the prime sets are implemented as hash-tables.

The algorithm is straightforward as well (Algorithm 2). It takes some set of triples (J), the current tricluster set $(\mathcal{T})$, and the dictionaries containing prime sets $(Primes)$ as input and outputs the modified versions of the tricluster set and dictionaries. The algorithm processes each triple (g, m, b) of J sequentially (line 1). At each iteration, the algorithm modifies the corresponding prime sets (lines 2–4).

Finally, it adds a new tricluster to the tricluster set. Note that this tricluster contains pointers to the corresponding prime sets (in the corresponding dictionaries) instead of the copies of the prime sets (line 5) which allows lowering the memory and access costs.

The algorithm is one-pass and its time and memory complexities are $O(|I|)$.

Duplicate elimination and selection patterns by user-specific constraints are done as post-processing to avoid patterns' loss. The time complexity of the basic post-processing is $O(|I|)$ and it does not require any additional memory.

The algorithm can be easily parallelised by splitting the subset of triples J into several subsets, processing each of them independently, and merging the resulting sets afterwards. This fact results in our previous MapReduce implementation [45].

Algorithm 2 Add function for the online algorithm for prime OAC-triclustering

Require: J is a set of triples;
    $\mathscr{T} = \{T = (*X, *Y, *Z)\}$ is a current set of triclusters;
    $PrimesOA, PrimesOC, PrimesAC$.
Ensure: $\mathscr{T} = \{T = (*X, *Y, *Z)\}$;
    $PrimesOA, PrimesOC, PrimesAC$.
 1: **for all** $(g, m, b) \in J$ **do**
 2:    $PrimesOA[g, m] := PrimesOA[g, m] \cup \{b\}$
 3:    $PrimesOC[g, b] := PrimesOC[g, b] \cup \{m\}$
 4:    $PrimesAC[m, b] := PrimesAC[m, b] \cup \{g\}$
 5:    $\mathscr{T} := \mathscr{T} \cup \{(\&PrimesAC[m, b], \&PrimesOC[g, b], \&PrimesOA[g, m])\}$
 6: **end for**

11.3 Triclustering Extensions

11.3.1 Multimodal Clustering

The direct extension of the prime object-attribute-condition triclustering is multimodal clustering for higher input relation arities. For the input polyadic context $\mathbb{K}_N = (A_1, A_2, \ldots, A_N, I \subseteq A_1 \times A_2 \times \cdots \times A_N)$ [40], we introduce the notion of *cumulus* for each input tuple $i = (e_1, e_2, \ldots, e_N) \in I$ and the corresponding entity e_k, where $k \in \{1, \ldots, N\}$ as follows:

$$cum(i, k) = \{e \mid (e_1, e_2, \ldots, e_{k-1}, e, e_{k+1}, \ldots, e_N) \in I\}.$$

The multimodal cluster generated by the tuple $i \in I$ is defined as follows:

$$((cum(i, 1), \ldots, cum(i, N)).$$

Those cumuli operators are similar to primes for pairs (or tuples) of sets Eq. 11.1:

$$cum(i, k) = (\{e_1\}, \{e_2\}, \ldots, \{e_{k-1}\}, \{e_{k+1}\}, \ldots, \{e_N\})'.$$

However, here, they are applied to the tuples of input relation rather than to pairs (tuples) of sets.

In a certain sense, cumuli accumulate all the elements of a fixed type that are related by I.

As its triadic version, multimodal clustering is not greater than the number of tuples in the input relation, whereas the complete set of polyadic concepts may be exponential w.r.t. the input size [40].

11.3.2 Many-Valued Triclustering

Another extension of prime OAC triclustering is many-valued triclustering [19] where each triple of the incidence relation of a triadic context is associated with a certain value from an arbitrary set W (by means of a valuation function $V : I \rightarrow W$). Therefore an input triadic context is changed to a many-valued triadic context $\mathbb{K}_V = (G, M, B, W, I, V)$, where for each triple (g, m, b) in ternary relation I between G, M, B, there is a unique $V(g, m, b) \in W$.

This definition contains the explicitly given valuation function V similar to the one in the definition of attribute-based information systems [34]. However, we could also extend the original definition of a many-valued context in line with [8]. In this case, a many-valued triadic context $\mathbb{K} = (G, M, B, W, J)$ consists of sets G, M, B, W and a quaternary relation between them $J \subseteq G \times M \times B \times W$, where the following holds: $(g, m, b, w) \in J$ and $(g, m, b, v) \in J$ imply $w = v$.

In what follows, we prefer the first variant since it simplifies further exposition.

For numeric values, the most common case is for $W = \mathbb{R}$. To mine triclusters in such context prime operators are modified into so-called δ-operators. For a generating triple $(\tilde{g}, \tilde{m}, \tilde{b}) \in I$ and some parameter δ:

$$(\tilde{m}, \tilde{b})^\delta = \left\{ g \mid (g, \tilde{m}, \tilde{b}) \in I \wedge |V(g, \tilde{m}, \tilde{b}) - V(\tilde{g}, \tilde{m}, \tilde{b})| \leq \delta \right\}$$

$$(\tilde{g}, \tilde{b})^\delta = \left\{ m \mid (\tilde{g}, m, \tilde{b}) \in I \wedge |V(\tilde{g}, m, \tilde{b}) - V(\tilde{g}, \tilde{m}, \tilde{b})| \leq \delta \right\}$$

$$(\tilde{g}, \tilde{m})^\delta = \left\{ b \mid (\tilde{g}, \tilde{m}, b) \in I \wedge |V(\tilde{g}, \tilde{m}, b) - V(\tilde{g}, \tilde{m}, \tilde{b})| \leq \delta \right\}$$

This definition can still be used to mine regular triclusters, if we set $W = \{0, 1\}$ and $\delta = 0$.

11.4 Implementations

11.4.1 Map-Reduce-Based Multimodal Clustering

We follow a three-stage approach here. On each stage, we sequentially run the map and reduce procedures: First map $\rightarrow$ First reduce $\rightarrow$ Second map $\rightarrow$ Second reduce $\rightarrow$ Third map $\rightarrow$ Third reduce. Each map/reduce procedure of a certain stage is executed in parallel on all the available nodes/clusters. How tasks are distributed among the nodes/clusters depends on the concrete MapReduce technology implementation (in our case, Apache Hadoop). Below, we describe the data flow between computational nodes and their processing.

(1) The first map (Algorithm 3) takes a set of input tuples. Each tuple $(e_1, e_2, \ldots, e_N)$ is transformed into N key-value pairs: $\langle (e_2, \ldots, e_N), e_1 \rangle, \langle (e_1, e_3, \ldots, e_N), e_2 \rangle, \ldots,$ $\langle (e_1, e_2, \ldots, e_{N-1}), e_N \rangle$. The resulting pairs are passed to the further step.

Algorithm 3 Distributed multimodal clustering: First Map

Require: I is a set of tuples of length N each
Ensure: $\langle subrelation, entity \rangle$ pairs.
1: **for all** $(e_1, e_2, \ldots, e_N) \in I$ **do**
2:    **for all** $k \in \{1, \ldots, N\}$ **do**
3:       $subrelation := (e_1, \ldots, e_{k-1}, e_{k+1}, \ldots, e_N)$
4:       **emit** $\langle subrelation, e_k \rangle$
5:    **end for**
6: **end for**

---

(2) The first reduce (Algorithm 4) receives all the accumulated values of each key. Thus, for each $(e_1, \ldots, e_N) \in I$ and the context entity type $k \in \{1, 2, \ldots, N\}$, we compute the cumulus $(e_1, \ldots, e_{k-1}, e_{k+1}, \ldots, e_N)'$. The values are passed to the next MapReduce stage with the key $(e_1, \ldots, e_{k-1}, e_{k+1}, \ldots, e_N)$.

Algorithm 4 Distributed multimodal clustering: First Reduce

Require: key-value pairs $\langle subrelation, entities \{e_k^1, \ldots, e_k^L\} \rangle$
Ensure: $\langle subrelation, cumulus \rangle$
1: $cumulus := \{\}$
2: **for all** $e_k \in \{e_k^1, \ldots, e_k^L\}$ **do**
3:    $cumulus := cumulus \cup \{e_k\}$
4:    **emit** $\langle subrelation, cumulus \rangle$
5: **end for**

---

(3) Second map (Algorithm 5). All the received keys are transformed into the original relations and passed to the second reduce procedure with unchanged values.

Algorithm 5 Distributed multimodal clustering: Second Map

Require: $\langle subrelation, cumulus \rangle$, where $subrelation = (e_1, \ldots, e_{k-1}, e_{k+1}, \ldots, e_N)$
Ensure: $\langle generating\_relation, cumulus \rangle$ pairs.
1: **for all** $e_k \in cumulus$ **do**
2:    $generating\_relation := (e_1, \ldots, e_{k-1}, e_k, e_{k+1}, \ldots, e_N)$
3:    **emit** $\langle generating\_relation, cumulus \rangle$
4: **end for**

---

(4) Second reduce (Algorithm 6). All the cumuli obtained for each input tuple of the original relation I are reduced to a single set. At this stage, we obtain all the original tuples and generated multimodal clusters. These clusters are presented as tuples of cumuli for respective entity types. All the obtained pairs $\langle generating\_relation, multimodal\_cluster \rangle$ are passed to the next stage.

(5) Third map (Algorithm 7). The task of the third MapReduce stage is duplicate elimination and filtration by density threshold. It is beneficial to im-

Algorithm 6 Distributed multimodal clustering: Second Reduce

Require: $\langle generating\_relation, cumuli\ \{A_1, A_2, \cdots, A_N\}\rangle$
Ensure: $\langle generating\_relation, multimodal\_cluster\rangle$ pairs
 1: $multimodal\_cluster := (A_1, A_2, \cdots, A_N)$
 2: **emit** $\langle generating\_relation, multimodal\_cluster\rangle$

plement within the reduce step, but to do so each obtained key-value pair $\langle generating\_relation,\ multimodal\_cluster\rangle$ should be passed further as follows $\langle multimodal\_cluster,\ generating\_relation\rangle$.

Algorithm 7 Distributed multimodal clustering: Third Map

Require: $\langle generating\_relation, multimodal\_cluster\rangle$
 1: **emit** $\langle multimodal\_cluster, generating\_relation\rangle$

(6) Third reduce (Algorithm 8). For each input multimodal cluster and its generating tuples, it is possible to directly compute density. All the unique clusters will be stored.

Algorithm 8 Distributed multimodal clustering: Third Reduce

Require: $\langle multimodal\_cluster, generating\_relations\ \{r_1, r_2 \ldots, r_M\}\rangle$
 1: **if** $\dfrac{|\{r_1, r_2 \ldots, r_M\}|}{vol(multimodal\_cluster)} \geq \theta$ **then**
 2:    **store** $\langle multimodal\_cluster\rangle$
 3: **end if**

The time and memory complexities are provided assuming that the worst-case scenario corresponds to the absolutely dense cuboid, i.e. polyadic context $\mathbb{K}_N = (A_1, A_2, \ldots, A_N, A_1 \times A_2 \times \cdots \times A_N)$. Thus, after careful analysis, the worst-case time complexity of the proposed three-stage algorithm is $O(|I| \sum_{j=1}^{N} |A_j|)$. Not surprisingly it has the same worst-case memory complexity since the stored and passed the maximal number of multimodal clusters is $|I|$, and the size of each of them is not greater $\sum_{j=1}^{N} |A_j|$. However, from an implementation point of view, since HDFS has default replication factor 3, those data elements are copied thrice to fulfil fault-tolerance.

11.4.2 Implementation Aspects and Used Technologies

The application[2] has been implemented in Java and as distributed computation framework we use Apache Hadoop.[3]

We have used many other technologies: Apache Maven (framework for automatic project assembling), Apache Commons (for work with extended Java collections), Jackson JSON (open-source library for the transformation of object-oriented representation of an object like tricluster to string), TypeTools (for real-time type resolution of inbound and outbound key-value pairs), etc.

To provide the reader with basic information on the most important classes for M/R implementation, let us shortly describe them below.

Entity. This is a basic abstraction for an element of a certain type. Each entity is defined by its type index from 0 to $n-1$, where n is the arity of the input formal context. An entity value is a string that needs to be kept during the program execution. This class inherits Writable interface for storing its objects in temporary and permanent Hadoop files. This is a mandatory requirement for all classes that pass or take their objects as keys and values of the map and reduce methods.

Tuple. This class implements a representation of relation. Each object of the class Tuple contains the list of objects of Entity class and the arity of its data given by its numeric value. This class implements interface WritableComparable<Tuple> to make it possible to use an object class Tuple as a key. The interface is similar to Writable, however, one needs to define comparison function to use in the key sorting phase.

Cumulus. This is an abstraction of cumulus, introduced earlier. It contains the list of string values and the index of a respective entity. It also implements the following interfaces: WritableComparable<Cumulus> for using cumulus as a key and Iterable<String> for iteration by its values.

FormalConcept. This is an abstraction of both formal concepts and multimodal clusters, it contains the list of cumuli and implements interface Writable.

The process-like M/R classes are summarised below.

FirstMapper, SecondMapper, ThirdMapper. These are the classes that extend class Mapper<> of the Hadoop MapReduce library by respective mapping function from Sect. 11.4.1.

FirstReducer, SecondReducer, ThirdReducer. These classes extend class Reducer<> of the Hadoop MapReduce library for fulfilling Algorithms 3,5,7.

TextCumulusInputFormat, TextCumulusOutputFormat. These classes implement reading and writing for objects of Cumulus class; they also inherit RecordReader and RecordWriter interfaces, respectively. They are required to exchange results between different MapReduce phases within one MapReduce program.

JobConfigurator. This is the class for setting configuration of a single MapReduce stage. It defines the classes of input/output/intermediate keys and values of the mapper and reducer as well as formats of input and output data.

---

[2] https://github.com/kostarion/multimodal-clustering-hadoop.

[3] https://hadoop.apache.org/.

App. This class is responsible for launching the application and chaining of M/R stages.

11.4.3 Parallel Many-Valued Triclustering

A generic algorithm for OAC triclustering is described below

Algorithm 9 General algorithm for OAC triclustering

Require: Context $\mathbb{K} = (G, M, B, I)$
Ensure: Set of triclusters $\mathscr{T}$
1:   $\mathscr{T} := \emptyset$
2: **for all** $(g, m, b) \in I$ **do**
3:     $oSet := applyPrimeOperator(m, b)$
4:     $aSet := applyPrimeOperator(g, b)$
5:     $cSet := applyPrimeOperator(g, m)$
6:     $tricluster := (oSet, aSet, cSet)$
7:     **if** $tricluster$ is valid **then**
8:       $Add(\mathscr{T}, tricluster)$
9:     **end if**
10: **end for**
11: **return** $\mathscr{T}$

To get a specific version of the algorithm one only needs to add an appropriate implementation of the prime operator and optional validity check. A tricluster mined from one triple does not depend on triclusters mined from other triples, so, in case of parallel implementation, each triple is processed in an individual thread.

We used δ-operators defined in Sect. 11.3.2, minimal density, and minimal cardinality (w.r.t. to every dimension) constraints [4].

11.5 Experiments

Two series of experiments have been conducted in order to test the application on the synthetic contexts and real-world datasets with a moderate and large number of triples in each. In each experiment, both versions of the OAC-triclustering algorithm have been used to extract triclusters from a given context. Only online and M/R versions of OAC-triclustering algorithm have managed to result in patterns for large contexts since the computation time of the compared algorithms was too high (>3000 s). To evaluate the runtime more carefully, for each context the average result of 5 runs of the algorithms has been recorded.

11.5.1 Datasets

Synthetic datasets. The following synthetic datasets were generated.

The dense context $\mathbb{K}_1 = (G, M, B, I)$, where $G = M = B = \{1, \ldots, 60\}$ and $I = G \times M \times B \setminus \{(g, m, b) \in I \mid g = m = b\}$. In total, 60^3 - 60 =215,940 triples.

The context of three non-overlapped cuboids $\mathbb{K}_2 = (G_1 \sqcup G_2 \sqcup G_3, M_1 \sqcup M_2 \sqcup M_3, B_1 \sqcup B_2 \sqcup B_3, I)$, where $I = (G_1 \times M_1 \times B_1) \cup (G_2 \times M_2 \times B_2) \cup (G_3 \times M_3 \times B_3)$. In total, $3 \cdot 50^3 = 375,000$ triples.

The context $\mathbb{K}_3 = (A_1, A_2, A_3, A_4, A_1 \times A_2 \times A_3 \times A_4)$ is a dense fourth dimensional cuboid with $|A_1| = |A_2| = |A_3| = |A_4| = 30$ containing $30^4 = 810,000$ tuples.

These tests have sense since in M/R setting due to the tuples can be (partially) repeated, e.g., because of M/R task failures on some nodes (i.e. restarting processing of some key-value pairs). Even though the third dataset does not result in $3^{min(|A_1|, |A_2|, |A_3|, |A_4|)}$ formal quadriconcepts , the worst-case for formal triconcepts generation in terms of the number of patterns, this is an example of the worst-case scenario for the reducers since the input has its maximal size w.r.t. to the size of A_i-s and the number of duplicates. In fact, our algorithm correctly assembles the only one quadricluster (A_1, A_2, A_3, A_4).

IMDB. This dataset consists of the 250 best movies from the Internet Movie Database based on user reviews.

The following triadic context is composed: the set of objects consists of movie names, the set of attributes (tags), the set of conditions (genres), and each triple of the ternary relation means that the given movie has the given genre and is assigned the given tag. In total, there are 3818 triples.

Movielens. The dataset contains 1,000,000 tuples that relate 6040 users, 3952 movies, ratings, and timestamps, where ratings are made on a 5-star scale [13].

Bibsonomy. Finally, a sample of the data of bibsonomy.org from ECML PKDD discovery challenge 2008 has been used.

This website allows users to share bookmarks and lists of literature and tag them. For the tests the following triadic context has been prepared: the set of objects consists of users, the set of attributes (tags), the set of conditions (bookmarks), and a triple of the ternary relation means that the given user has assigned the given tag to the given bookmark.

Table 11.2 contains the summary of IMDB and Bibsonomy contexts.

Context	$|G|$	$|M|$	$|B|$	# triples	Density
IMDB	250	795	22	3818	0.00087
Bibsonomy	2337	67,464	28,920	816,197	$1.8 \cdot 10^{-7}$

Table 11.2: Tricontexts based of real data systems for the experiments

Due to a number of requests, we would like to show small excerpts of the input data we used (in this section) and a few examples of the outputted patterns (in the next section).

Input Data Example. The input file for the Top-250 IMDB dataset comprises triples in lines with tab characters as separators.

One Flew Over the Cuckoo's Nest (1975)   Nurse   Drama
One Flew Over the Cuckoo's Nest (1975)   Patient   Drama
One Flew Over the Cuckoo's Nest (1975)   Asylum   Drama
One Flew Over the Cuckoo's Nest (1975)   Rebel   Drama
One Flew Over the Cuckoo's Nest (1975)   Basketball   Drama
Star Wars V: The Empire Strikes Back (1980)   Princess   Action
Star Wars V: The Empire Strikes Back (1980)   Princess   Adventure
Star Wars V: The Empire Strikes Back (1980)   Princess   Sci-Fi
. . .

11.5.2 Results

The experiments have been conducted on the computer Intel®Core(TM) i5-2450M CPU @ 2.50GHz, 4Gb RAM (typical commodity hardware) in the emulation mode, when Hadoop cluster contains only one node and operates locally and sequentially. By time execution results one can estimate the performance in a real distributed environment assuming that each node workload is (roughly) the same.

Method	IMDB	MovieLens100k	$\mathbb{K}_1$	$\mathbb{K}_2$	$\mathbb{K}_3$
Online OAC prime clustering	368	16,298	96,990	185,072	643,978
MapReduce multimodal clustering	7124	14,582	37,572	61,367	102,699

Table 11.3: Three-stage MapReduce multimodal clustering time, ms

In Tables 11.3 and 11.4 we summarise the results of performed tests. It is clear that on average our application has a smaller execution time than its competitor, the online version of OAC-triclustering. If we compare the implemented program with its original online version, the results are worse for the not that big and sparse dataset as IMDB. It is the consequence of the fact that the application architecture aimed at processing large amounts of data; in particular, it is implemented in three stages with time-consuming communication. Launching and stopping Apache Hadoop, data writing, and passing between Map and Reduce steps in both stages requires substantial time, that is why for not that big datasets when execution time is comparable with time for infrastructure management, time performance is not

Dataset	Online OAC Prime	M/R total	MapReduce stages			# clusters
			1st	2nd	3rd	
MovieLens100k	89,931	16,348	8724	5292	2332	89,932
MovieLens250k	225,242	42,708	10,075	20,338	12,295	225,251
MovieLens500k	461,198	94,701	15,016	46,300	33,384	461,238
MovieLens1M	958,345	217,694	28,027	114,221	74,446	942,757
Bibsonomy (≈800k triples)	>6 h	3,651,072 (≈1 h)	19,117	1,972,135	1,659,820	486,221

Table 11.4: Three-stage MapReduce multimodal clustering time, ms

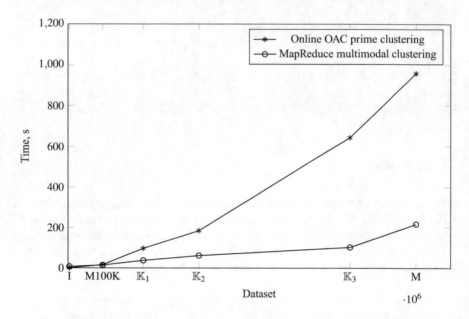

Fig. 11.2: Performance curves for six datasets: I stands for IMDB dataset with 3818 triples, M100K—MovieLens dataset with 100K tuples, M—MovieLens dataset with 1M tuples

perfect. However, with data size increase the relative performance is growing up to five-six times (see Fig. 11.2). Thus, the last test for BibSonomy data has been successfully passed, but the competitor was not able to finish it within one hour. As for the M/R stages, the most time-consuming phases are the 2nd and 3rd stages.

Output triclusters example. The output triclusters are stored in a format, similar to JSON: the sets of entities (modalities) are given in curly brackets separated with commas. Both sets of entities and triclusters start with a new line.

Thus, for the modality of movies, in case of IMDB data, we can see two remaining modalities as keywords and genres.

```
{
{Apocalypse Now (1979), Forrest Gump (1994),
Full Metal Jacket (1987), Platoon (1986)}
{Vietnam}
{Drama, Action}
}
{
{Toy Story (1995), Toy Story 2 (1999)}
{Toy, Friend}
{Animation, Adventure, Comedy, Family, Fantasy}
}
{
{Star Wars: Episode V - The Empire Strikes Back (1980),
WALL-E (2008), Toy Story 2 (1999)}
{Rescue}
{Animation, Adventure}
}
{
{Into the Wild (2007), The Gold Rush (1925)}
{Love, Alaska}
{Adventure}
}
...
```

11.6 Experiments with Parallelisation

In the additional set of experiments, we investigated parallelisation as another way of improving the performance of our triclustering algorithms. Modern computers support multi-threading and have several processing cores. However one needs special instructions and thread-safe data structures to produce an efficient parallel algorithm.

We used a dataset on semantic tri-frames featured in [39]. These triples represent semantic frames extracted from FrameNet1.7 [1] and each triple is accompanied by a frequency from DepCC dataset [33]. The total count of triples reached 100 thousand. The data was processed with the NOAC algorithm [4] implemented in C# .NetFramework 4.5 [5] and modified for parallel computation with Parallel library, namely each triple from the context is processed in a separate thread. We ran two series of experiments with different algorithm parameters and measured execution time against the number of processed triples and also provided the number of extracted triclusters. In Table 11.5, for example, the record NOAC(100, 0.8, 2) 1k means that $\delta = 100$, $\rho_{min} = 0.8$, $minsup = 2$ (for each dimension) and 1000 triples are being processed. All additional experiments were conducted on an Intel®Core(TM) i7-8750H CPU @ 2.20GHz, 16Gb RAM.

Experiment	Time, ms (regular)	Time, ms (parallel)	# Triclusters
NOAC(100, 0.8, 2) 1k	109	117	0
NOAC(100, 0.8, 2) 10k	5025	3642	3
NOAC(100, 0.8, 2) 20k	16,825	11,759	20
NOAC(100, 0.8, 2) 30k	33,067	22,519	52
NOAC(100, 0.8, 2) 40k	56,878	36,994	92
NOAC(100, 0.8, 2) 50k	80,095	52,322	116
NOAC(100, 0.8, 2) 60k	102,092	67,748	145
NOAC(100, 0.8, 2) 70k	133,974	87,956	160
NOAC(100, 0.8, 2) 80k	175,597	110,044	201
NOAC(100, 0.8, 2) 90k	223,932	135,268	223
NOAC(100, 0.8, 2) 100k	268,021	157,073	254
NOAC(100, 0.5, 0) 1k	110	169	803
NOAC(100, 0.5, 0) 10k	5121	3681	4942
NOAC(100, 0.5, 0) 50k	82,130	52,558	14,214
NOAC(100, 0.5, 0) 100k	268,128	159,333	23,134

Table 11.5: NOAC. Regular ans parallel version

These experiments show that performance noticeably benefits from parallelisation (see Fig. 11.3). The execution time of the parallel algorithm is on average 35% lower. Another interesting outcome is that execution time does not depend on the algorithm parameters, which only change the number of extracted triclusters.

11.7 Conclusion

In this paper, we have presented a map-reduce version of the multimodal clustering algorithm, which extends triclustering approaches and copes with bottlenecks of the earlier M/R triclustering version [45]. We have shown that the proposed algorithm is efficient from both theoretical and practical points of view. This is a variant of map-reduce based algorithm where the reducer exploits composite keys directly (see also Appendix section [45]). However, despite the step towards Big Data technologies, a proper comparison of the proposed multimodal clustering and noise-tolerant patterns in n-ary relations by DataPeeler and its descendants [3] is not yet conducted (including MapReduce setting).

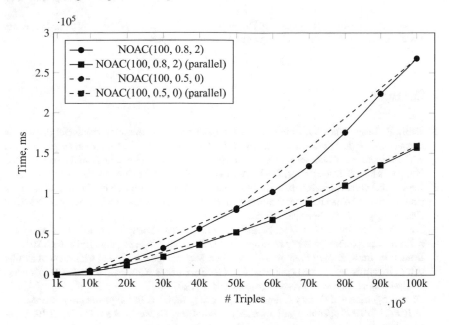

Fig. 11.3: Performance curves for parallelisation experiments

Two of the most challenging problems in OAC-triclustering are approximate triclustering density estimation (e.g., employing the Monte Carlo approach) and duplicate elimination (the same triclusters can be generated by different generating triples). In MapReduce setting, both procedures are implemented as separate MapReduce stages, while for online triclustering generation it is better to avoid duplicate generation, and the tricluster density should be computed both approximately to maintain appropriate time-complexity and iteratively with minimal updates for each incoming triple.

Further development of the proposed triclustering methods for large datasets is possible with Apache Spark[4] [43].

We also examined parallelisation possibilities (namely, multi-thread and milticore advantages in modern C#) of the numeric triclustering algorithm NOAC and showed that it may significantly improve performance.

Acknowledgements The study was implemented in the framework of the Basic Research Program at the National Research University Higher School of Economics (Sects. 11.2 and 11.5), and funded by the Russian Academic Excellence Project '5-100'. The second author was also supported by the Russian Science Foundation (Sects. 11.1, 11.3, and 11.4) under grant 17-11-01294. The authors would like to thank Dmitry Gnatyshak and Sergey Zudin for their earlier work on incremental and MapReduce-based triclustering implementations, respectively, anonymous reviewers as well

---

[4] https://spark.apache.org/.

as Yuri Kudriavtsev from PM-Square and Dominik Slezak from Infobright and Warsaw University for their encouragement given to our studies of M/R technologies.

References

1. Collin F. Baker, Charles J. Fillmore, and John B. Lowe. The Berkeley FrameNet Project. In *Proceedings of the 36th Annual Meeting of the Association for Computational Linguistics and 17th International Conference on Computational Linguistics - Volume 1*, ACL '98/COLING '98, pages 86–90, Stroudsburg, PA, USA, 1998. Association for Computational Linguistics.
2. Besson, J., Pensa, R.G., Robardet, C., Boulicaut, J.F.: Constraint-based mining of fault-tolerant patterns from boolean data. In: International Workshop on Knowledge Discovery in Inductive Databases. pp. 55–71. Springer (2005)
3. Loïc Cerf, Jérémy Besson, Kim-Ngan Nguyen, and Jean-François Boulicaut. Closed and noise-tolerant patterns in n-ary relations. *Data Min. Knowl. Discov.*, 26(3):574–619, 2013.
4. Dmitrii Egurnov, Dmitry Ignatov, and Engelbert Mephu Nguifo. Mining triclusters of similar values in triadic real-valued contexts. In *14th International Conference on Formal Concept Analysis-Supplementary Proceedings*, pages 31–47, 2017.
5. Dmitrii Egurnov and Dmitry I. Ignatov. Triclustering toolbox. In *Supplementary Proceedings of ICFCA 2019 Conference and Workshops, Frankfurt, Germany, June 25–28, 2019*, pages 65–69, 2019.
6. Kemal Eren, Mehmet Deveci, Onur Kucuktunc, and Catalyurek, Umit V. A comparative analysis of biclustering algorithms for gene expression data. *Briefings in Bioinform.*, 2012.
7. Bernhard Ganter, Peter A. Grigoriev, Sergei O. Kuznetsov, and Mikhail V. Samokhin. Concept-based data mining with scaled labeled graphs. In *ICCS*, pages 94–108, 2004.
8. Bernhard Ganter and Rudolf Wille. *Formal Concept Analysis: Mathematical Foundations*. Springer-Verlag New York, Inc., Secaucus, NJ, USA, 1st edition, 1999.
9. Elisabeth Georgii, Koji Tsuda, and Bernhard Schölkopf. Multi-way set enumeration in weight tensors. *Machine Learning*, 82(2):123–155, 2011.
10. Dmitry V. Gnatyshak, Dmitry I. Ignatov, and Sergei O. Kuznetsov. From triadic FCA to triclustering: Experimental comparison of some triclustering algorithms. In *CLA*, pages 249–260, 2013.
11. Dmitry V. Gnatyshak, Dmitry I. Ignatov, Sergei O. Kuznetsov, and Lhouari Nourine. A one-pass triclustering approach: Is there any room for big data? In *CLA 2014*, 2014.
12. Dmitry V. Gnatyshak, Dmitry I. Ignatov, Alexander V. Semenov, and Jonas Poelmans. Gaining insight in social networks with biclustering and triclustering. In *BIR*, volume 128 of *Lecture Notes in Business Information Processing*, pages 162–171. Springer, 2012.
13. F. Maxwell Harper and Joseph A. Konstan. The MovieLens datasets: History and context. *TiiS*, 5(4):19:1–19:19, 2016.
14. Rui Henriques and Sara C. Madeira. Triclustering algorithms for three-dimensional data analysis: A comprehensive survey. *ACM Comput. Surv.*, 51(5):95:1–95:43, September 2018.
15. Dmitry I. Ignatov, Dmitry V. Gnatyshak, Sergei O. Kuznetsov, and Boris Mirkin. Triadic formal concept analysis and triclustering: searching for optimal patterns. *Machine Learning*, pages 1–32, 2015.
16. Dmitry I. Ignatov, Sergei O. Kuznetsov, and Jonas Poelmans. Concept-based biclustering for internet advertisement. In *ICDM Workshops*, pages 123–130. IEEE Computer Society, 2012.
17. Dmitry I. Ignatov, Sergei O. Kuznetsov, Jonas Poelmans, and Leonid E. Zhukov. Can triconcepts become triclusters? *International Journal of General Systems*, 42(6):572–593, 2013.

18. Dmitry I. Ignatov, Elena Nenova, Natalia Konstantinova, and Andrey V. Konstantinov. Boolean Matrix Factorisation for Collaborative Filtering: An FCA-Based Approach. In *AIMSA 2014, Varna, Bulgaria, Proceedings*, volume LNCS 8722, pages 47–58, 2014.

19. Dmitry I. Ignatov, Dmitry Tochilkin, and Dmitry Egurnov. Multimodal clustering of Boolean tensors on MapReduce: Experiments revisited. In *Supplementary Proceedings of ICFCA 2019 Conference and Workshops, Frankfurt, Germany, June 25–28, 2019*, pages 137–151, 2019.

20. Mohamed Nader Jelassi, Sadok Ben Yahia, and Engelbert Mephu Nguifo. A personalized recommender system based on users' information in folksonomies. In Leslie Carr and et al., editors, *WWW (Companion Volume)*, pages 1215–1224. ACM, 2013.

21. Mehdi Kaytoue, Sergei O. Kuznetsov, Juraj Macko, and Amedeo Napoli. Biclustering meets triadic concept analysis. *Ann. Math. Artif. Intell.*, 70(1–2):55–79, 2014.

22. Mehdi Kaytoue, Sergei O. Kuznetsov, Amedeo Napoli, and Sébastien Duplessis. Mining gene expression data with pattern structures in formal concept analysis. *Inf. Sci.*, 181(10):1989–2001, 2011.

23. Petr Krajca and Vilem Vychodil. Distributed algorithm for computing formal concepts using map-reduce framework. In *N. Adams et al. (Eds.): IDA 2009*, volume LNCS 5772, pages 333–344, 2009.

24. Sergey Kuznecov and Yury Kudryavcev. Applying map-reduce paradigm for parallel closed cube computation. In *1st Int. Conf. on Advances in Databases, Knowledge, and Data Applications, DBKDS 2009*, pages 62–67, 2009.

25. Lehmann, F., Wille, R.: A Triadic Approach to Formal Concept Analysis. In: Proceedings of the Third International Conference on Conceptual Structures: Applications, Implementation and Theory. pp. 32–43 (1995)

26. Ao Li and David Tuck. An effective tri-clustering algorithm combining expression data with gene regulation information. *Gene regul. and syst. biol.*, 3:49–64, 2009.

27. Sara C. Madeira and Arlindo L. Oliveira. Biclustering algorithms for biological data analysis: A survey. *IEEE/ACM Trans. Comput. Biology Bioinform.*, 1(1):24–45, 2004.

28. Saskia Metzler and Pauli Miettinen. Clustering Boolean tensors. *Data Min. Knowl. Discov.*, 29(5):1343–1373, 2015.

29. Boris Mirkin. *Mathematical Classification and Clustering*. Kluwer, Dordrecht, 1996.

30. Boris G. Mirkin and Andrey V. Kramarenko. Approximate bicluster and tricluster boxes in the analysis of binary data. In Sergei O. Kuznetsov and et al., editors, *RSFDGrC 2011*, volume 6743 of *Lecture Notes in Computer Science*, pages 248–256. Springer, 2011.

31. Rokia Missaoui and Léonard Kwuida. Mining triadic association rules from ternary relations. In *Formal Concept Analysis - 9th International Conference, ICFCA 2011, Nicosia, Cyprus, May 2–6, 2011. Proceedings*, pages 204–218, 2011.

32. Alexandros Nanopoulos, Dimitrios Rafailidis, Panagiotis Symeonidis, and Yannis Manolopoulos. Musicbox: Personalized music recommendation based on cubic analysis of social tags. *IEEE Transactions on Audio, Speech & Language Processing*, 18(2):407–412, 2010.

33. Alexander Panchenko, Eugen Ruppert, Stefano Faralli, Simone Paolo Ponzetto, and Chris Biemann. Building a Web-Scale Dependency-Parsed Corpus from CommonCrawl. In Nicoletta Calzolari (Conference chair), Khalid Choukri, Christopher Cieri, Thierry Declerck, Sara Goggi, Koiti Hasida, Hitoshi Isahara, Bente Maegaard, Joseph Mariani, Hélène Mazo, Asuncion Moreno, Jan Odijk, Stelios Piperidis, and Takenobu Tokunaga, editors, *Proceedings of the Eleventh International Conference on Language Resources and Evaluation (LREC 2018)*, Miyazaki, Japan, May 7–12, 2018 2018. European Language Resources Association (ELRA).

34. Z. Pawlak. Information systems – theoretical foundations. *Information Systems*, 6(3):205 – 218, 1981.

35. Anand Rajaraman, Jure Leskovec, and Jeffrey D. Ullman. *Mining of Massive Datasets*, chapter MapReduce and the New Software Stack, pages 19–70. Cambridge University Press, England, Cambridge, 2013.

36. Nicole Schweikardt. One-pass algorithm. In *Encyclopedia of Database Systems, Second Edition*. 2018.
37. Kijung Shin, Bryan Hooi, and Christos Faloutsos. Fast, accurate, and flexible algorithms for dense subtensor mining. *TKDD*, 12(3):28:1–28:30, 2018.
38. Eirini Spyropoulou, Tijl De Bie, and Mario Boley. Interesting pattern mining in multi-relational data. *Data Mining and Knowledge Discovery*, 28(3):808–849, 2014.
39. Dmitry Ustalov, Alexander Panchenko, Andrey Kutuzov, Chris Biemann, and Simone Paolo Ponzetto. Unsupervised semantic frame induction using triclustering. In Iryna Gurevych and Yusuke Miyao, editors, *Proceedings of the 56th Annual Meeting of the Association for Computational Linguistics, ACL 2018, Melbourne, Australia, July 15–20, 2018, Volume 2: Short Papers*, pages 55–62. Association for Computational Linguistics, 2018.
40. George Voutsadakis. Polyadic concept analysis. *Order*, 19(3):295–304, 2002.
41. Rudolf Wille. Restructuring lattice theory: An approach based on hierarchies of concepts. In Ivan Rival, editor, *Ordered Sets*, volume 83 of *NATO Advanced Study Institutes Series*, pages 445–470. Springer Netherlands, 1982.
42. Biao Xu, Ruairı de Frein, Eric Robson, and Micheal O Foghlu. Distributed Formal Concept Analysis algorithms based on an iterative MapReduce framework. In F. Domenach, D.I. Ignatov, and J. Poelmans, editors, *ICFCA 2012*, volume LNAI 7278, pages 292–308, 2012.
43. Matei Zaharia, Reynold S. Xin, Patrick Wendell, Tathagata Das, Michael Armbrust, Ankur Dave, Xiangrui Meng, Josh Rosen, Shivaram Venkataraman, Michael J. Franklin, Ali Ghodsi, Joseph Gonzalez, Scott Shenker, and Ion Stoica. Apache Spark: A unified engine for big data processing. *Commun. ACM*, 59(11):56–65, October 2016.
44. Lizhuang Zhao and Mohammed Javeed Zaki. Tricluster: An effective algorithm for mining coherent clusters in 3d microarray data. In *SIGMOD 2005 Conference*, pages 694–705, 2005.
45. Sergey Zudin, Dmitry V. Gnatyshak, and Dmitry I. Ignatov. Putting OAC-triclustering on MapReduce. In Sadok Ben Yahia and Jan Konecny, editors, *Proceedings of the Twelfth International Conference on Concept Lattices and Their Applications, Clermont-Ferrand, France, October 13–16, 2015.*, volume 1466 of *CEUR Workshop Proceedings*, pages 47–58. CEUR-WS.org, 2015.

Index

Printed in the United States
by Baker & Taylor Publisher Services